ESSENTIALS OF ECOLOGY

Edited by
Daniel Dela Torre

ArclerPress

Essentials of Ecology

Edited by Daniel Dela Torre

Edition 2018

ArclerPress

Published by
Arcler Press LLC
708 3rd Avenue, 6th Floor
New York
NY 10017
United States of America
www.arclerpress.com

ISBN: 978-1-68094-619-2 (Hardback)
Printed in United States of America
Library of Congress Control Number: 2017933215

About the Editor

Daniel Dela Torre

Daniel completed his MSc in Environmental Science from the University of the Philippines Diliman in 2016. His primary interests are ecological monitoring using remote sensing and the effects of climate variability on vegetation. He is engaged in the University of the Philippines as a Senior Science Research Specialist working on LiDAR mapping of agricultural resources.

Table of Contents

2 STUDY OF ALLOMETRY

3 ECOLOGICAL ECONOMICS

4 BALANCE OF NATURE

5 CONSTRUCTAL LAW

6 ECOLOGICAL FORECASTING

7 ECOLOGICAL STOICHIOMETRY

8 FISHERIES FISH ECOLOGY AND CLIMATE CHANGE

11 REGIME SHIFT

12 THEORETICAL ECOLOGY

13 ECOLOGICAL EFFECTS OF BIODIVERSITY

List of Abbreviations

AHB	Africanized honey bees
ASS	Alternative stable states
EIA	Environmental impact assessment
EPA	Environmental Protection Agency
EEZ	Exclusive economic zones
CBA	Cost benefit analysis
FAO	Cost benefit analysis
GMF	Genetically modified food
Gt	Gigatonnes
ESEM	Earth systems engineering and management
EE	Ecological economics
EFA	Ecological footprint analysis
EF	Ecological footprint
EwE	Ecopath with Ecosim
EROEI	Energy return on energy invested
ITQ	Individual transferable quotas
IE	Industrial ecology
IOA	Input-output analysis
LDG	Latitudinal diversity gradient
LCA	Life-cycle assessment
LCC	Life cycle costing
LTER	Long-term ecological research

NPP	Net primary productivity
NCE	Neoclassical economics
PCR	Polymerase chain reaction
UNCLOS	United Nations Convention on the Law of the Sea
SFA	Substance flow analysis

Preface

The natural world is made up of living and non-living things. Interactions between these components especially in resource sharing, allocation and use, have an important function in the environment. To understand interrelationships existing in our environment we need to look into ecology. Ecology is the study of relationships between living organisms, including humans, and their physical environment; it seeks to understand the vital connections between plants and animals and the world around them. Ecology also provides information about the benefits of ecosystems and how we can use Earth's resources in sustainable ways to supply the needs of future generations.

Ecology elucidates the ways wherein people and nature depend on one another. Specialized fields in ecology provide us with information to better understand our world and also help us improve our environment, manage our natural resources, and protect human health. Predictions on the human effects on the environment is borne out of understanding how ecological systems work. Studying how different species rely on one another can also aid in protecting critical populations of species. Lack of understanding of these relationships have led to reduction of species numbers and even to extinction in the past.

Human populations can also benefit from the natural services that ecosystems provide. Ecologists have discovered that marshes and wetlands filter toxins and other impurities from water. Leaving some of these filtering ecosystems intact can reduce the burden on water treatment plants that have been built to perform the same service. By using natural filtering systems, we have the option to build less treatment plants. Harnessing these natural services has been a key point in global initiatives to mitigate environmental degradation.

This book provides opportunities for students to study, practice, and develop quantitative and analytical skills in learning about ecology.

Content Coverage

Chapter One presents an overview of the scientific analysis and study of interactions among organisms and their environment.

Chapters Two and *Three* are biological scaling (allometry) and ecological economics. Allometry is the study of the relationship of body size to shape, anatomy, physiology and finally behavior.

Chapter Four proposes the balance of nature, a theory proposing that ecological systems are usually in a stable equilibrium, which is to say that a small change in some particular parameter will be corrected by some negative feedback that will bring the parameter back to its original state at the point of balance in relation to the rest of the system.

Chapters Five and *Six* present constructal law and ecological forecasting, separately. Ecological forecasting uses knowledge of physics, ecology and physiology to predict how ecosystems will change in the future in response to environmental factors such as climate change.

Chapter Seven covers ecological stoichiometry considers how the balance of energy and elements affects and is affected by organisms and their interactions in ecosystems.

Chapters Eight and *Nine* focus on fish ecology, and industrial ecology. Industrial ecology (IE) is the study of material and energy flows through industrial systems.

Chapters Ten and *Eleven* include all living and non-living things occurring naturally, and regime shifts, respectively.

Chapters Twelve and *Thirteen* are theoretical ecology and ecological effects of biodiversity, correspondingly. Theoretical ecology is the use of conceptual, mathematical and computational methods to address ecological problems whereas ecological effects of biodiversity in turn are affected by both climate change through enhanced greenhouse gases, aerosols and loss of land cover, and biological diversity, causing a rapid loss of ecosystems and extinctions of species and local populations.

CHAPTER 1

Ecology: An overview

Chapter Contents

- **Introduction**
- **Levels, Scope, and Scale of Organization**
- **Ecological Complexity**
- **Relation to Evolution**
- **Human Ecology**
- **Relation to the Environment**
- **History**

Ecology is not just about species-rich forests, pristine wilderness, or scenic vistas, though. Have you, for instance, ever found cockroaches living under your bed, mold growing in your shower, or even fungus creeping in between your toes? If so, then you have seen equally valid examples of ecology in action.

INTRODUCTION

Ecology is the scientific analysis and study of interactions among organisms and their environment. It is an interdisciplinary field that includes biology, geography, and Earth science. Ecology includes the study of interactions organisms have with each other, other organisms, and with abiotic components of their environment. Topics of interest to ecologists include the diversity, distribution, amount (biomass), and number (population) of particular organisms, as well as cooperation and competition between organisms, both within and among ecosystems. Ecosystems are composed of dynamically interacting parts including organisms, the communities they make up, and the non-living components of their environment. Ecosystem processes, such as primary production, pedogenesis, nutrient cycling, and various niche construction activities, regulate the flux of energy and matter through an environment. These processes are sustained by organisms with specific life history traits, and the variety of organisms is called biodiversity. Biodiversity, which refers to the varieties of species, genes, and ecosystems, enhances certain ecosystem services.

Ecology is closely related to allied fields such as evolutionary biology, genetics and ethology. An important focus for ecologists is to improve the understanding of how biodiversity affects ecological function. Ecologists seek to explain:

- Life processes, interactions, and adaptations
- The movement of materials and energy through living communities
- The successional development of ecosystems
- The abundance and distribution of organisms and biodiversity in the context of the environment.

Ecology is a human science as well. There are many practical applications of ecology in conservation biology, wetland management, natural resource management (agroecology, agriculture, forestry, agroforestry, fisheries), city planning (urban ecology), community health, economics, basic and applied science, and human social interaction (human ecology). For example, the *Circles of Sustainability* approach treats ecology as more than the environment 'out there, where the role of humans is integrated, rather than separated, from the study of the natural world. Organisms, including humans, and resources compose ecosystems which, in turn, maintain biophysical feedback mechanisms that moderate processes acting on living (biotic) and non-living (abiotic) components of the planet. Ecosystems sustain life-supporting functions and produce natural capital like biomass production (food, fuel, fiber, and medicine), the regulation of climate, global biogeochemical cycles, water filtration, soil formation, erosion control, flood protection, and many other natural features of scientific, historical, economic, or intrinsic value.

The word ecology was coined in 1866 by the German scientist Ernst Haeckel (1834–1919). Ecological thought was derivative of established trends in philosophy, particularly from ethics and politics. Ancient Greek philosophers such as Hippocrates and Aristotle laid the foundations of ecology in their studies on natural history. Modern ecology became a much more rigorous science in the late 19th century. Evolutionary concepts relating to adaptation and natural selection became the cornerstones of modern ecological theory.

LEVELS, SCOPE, AND SCALE OF ORGANIZATION

The scope of ecology contains a wide array of interacting levels of organization spanning micro-level (e.g., cells) to a planetary scale (e.g., biosphere) phenomena. Ecosystems, for example, contain abiotic resources and interacting life forms (i.e., individual organisms that aggregate into populations which aggregate into distinct ecological communities). Ecosystems are dynamic, they do not always follow a linear successional path, but they are always changing, sometimes rapidly and sometimes so slowly that it can take thousands of years for ecological processes to bring about certain successional stages of a forest. An ecosystem's area can vary greatly, from tiny to vast. A single tree is of little consequence to the classification of a forest ecosystem, but critically relevant to organisms living in and on it. Several generations of an aphid population can exist over the lifespan of a single leaf. Each of those aphids, in turn, support diverse bacterial communities. The nature of connections in ecological communities cannot be explained by knowing the details of each species in isolation, because the emergent pattern is neither revealed nor predicted until the ecosystem is studied as an integrated whole. Some ecological principles, however, do exhibit collective properties where the sum of the components explain the properties of the whole, such as birth rates of a population being equal to the sum of individual births over a designated time frame.

Hierarchical Ecology

System behaviors must first be arrayed into different levels of organization. Behaviors corresponding to higher levels occur at slow rates. Conversely, lower organizational levels exhibit rapid rates. For example, individual tree leaves respond rapidly to momentary changes in light intensity, CO_2 concentration, and the like. The growth of the tree responds more slowly and integrates these short-term changes.

ONeill et al. (1986)

The scale of ecological dynamics can operate like a closed system, such as aphids migrating on a single tree, while at the same time remain open with regard to broader scale influences, such as atmosphere or climate. Hence, ecologists classify ecosystems hierarchically by analyzing data collected from finer scale units, such as vegetation associations, climate, and soil types, and integrate this information to identify emergent patterns of uniform organization and processes that operate on local to regional, landscape, and chronological scales.

To structure the study of ecology into a conceptually manageable framework, the biological world is organized into a nested hierarchy of increasing scale as follows:

genes --> cells --> tissues --> organs --> organisms --> species --> populations --> communities --> ecosystems --> biomes --> biosphere.

This framework forms a panarchy and exhibits non-linear behaviors, meaning that "effect and cause are disproportionate, so that small changes to critical variables, such as the number of nitrogen fixers, can lead to disproportionate, perhaps irreversible, changes in the system properties.

Biodiversity

Biodiversity refers to the variety of life and its processes. It includes the variety of living organisms, the genetic differences among them, the communities and ecosystems in which they occur, and the ecological and evolutionary processes that keep them functioning, yet ever changing and adapting.

Biodiversity (abbreviated from "biological diversity") describes the diversity of life from genes to ecosystems and spans every level of biological organization. The term has several interpretations, and there are many ways to index, measure, characterize, and represent its complex organization. Biodiversity includes species diversity, ecosystem diversity, and genetic diversity and scientists are interested in the way that this diversity affects the complex ecological processes operating at and among these respective levels.

Biodiversity plays an important role in ecosystem services which by definition maintain and improve human quality of life. Preventing species extinctions is one way to preserve biodiversity and that goal rests on techniques that preserve species genetic diversity, habitat integrityt and migration ability. Conservation priorities and management techniques require different approaches and considerations to address the full ecological scope of biodiversity. Natural capital that supports populations is critical for maintaining ecosystem services and species migration (e.g., riverine fish runs and avian insect control) has been implicated as one mechanism by which those service losses are experienced. An understanding of biodiversity has practical applications for species and ecosystem-level conservation planners as they make management recommendations to consulting firms, governments, and industry.

Habitat

The habitat of a species describes the environment over which a species is known to occur and the type of community that is formed as a result. More specifically, "habitats can be defined as regions in environmental space that are composed of multiple dimensions, each representing a biotic or abiotic environmental variable; that is, any component or characteristic of the environment related directly (e.g. forage biomass and quality) or indirectly (e.g. elevation) to the use of a location by the animal." For example, a habitat might be an aquatic or terrestrial environment that can be further categorized as a montane or alpine ecosystem.

Habitat shifts provide important evidence of competition in nature where one population changes relative to the habitats that most other individuals of the species occupy. For example, one population of a species of tropical lizards (*Tropidurus hispidus*) has a flattened body relative to the main populations that live in open savanna. The population that lives in an isolated rock outcrop hides in crevasses where its flattened body offers a selective advantage. Habitat shifts also occur in the developmental life history of amphibians, and in insects that transition from aquatic to terrestrial habitats. Biotope and habitat are sometimes used interchangeably, but the former applies to a community's environment, whereas the latter applies to a species' environment.

Additionally, some species are *ecosystem engineers*, altering the environment within a localized region. For instance, beavers manage water levels by building dams which improves their habitat in a landscape.

Biodiversity of a coral reef. Corals adapt to and modify their environment by forming calcium carbonate skeletons. This provides growing conditions for future generations and forms a habitat for many other species.

Niche

Definitions of the niche date back to 1917, but G. Evelyn Hutchinson made conceptual advances in 1957 by introducing a widely adopted definition: "the set of biotic and abiotic conditions in which a species is able to persist and maintain stable population sizes. The ecological niche is a central concept in the ecology of organisms and is subdivided into the *fundamental* and the *realized* niche. The fundamental niche is the set of environmental conditions under which a species is able to persist. The realized niche is the set of environmental plus ecological conditions under which a species persists. The Hutchinsonian niche is defined more technically as a "Euclidean hyperspace whose *dimensions* are defined as environmental variables and whose *size* is a function of the number of values that the environmental values may assume for which an organism has *positive fitness*."

Termite mounds with varied heights of chimneys regulate gas exchange, temperature and other environmental parameters that are needed to sustain the internal physiology of the entire colony.

Biogeographical patterns and range distributions are explained or predicted through knowledge of a species' traits and niche requirements. Species have functional traits that are uniquely adapted to the ecological niche. A trait is a measurable property, phenotype, or characteristic of an organism that may influence its survival. Genes play an important role in the interplay of development and environmental expression of traits. Resident species evolve traits that are fitted to the selection pressures of their local environment. This tends to afford them a competitive advantage and discourages similarly adapted species from having an overlapping geographic range. The competitive exclusion principle states that two species cannot coexist indefinitely by living off the same limiting resource; one will always out-compete the other. When similarly adapted species overlap geographically, closer inspection reveals subtle ecological differences in their habitat or dietary requirements. Some models and empirical studies, however, suggest that disturbances can stabilize the co-evolution and shared niche occupancy of similar species inhabiting species-rich communities. The habitat plus the niche is called the ecotope, which is defined as the full range of environmental and biological variables affecting an entire species.

Niche Construction

Organisms are subject to environmental pressures, but they also modify their habitats. The regulatory feedback between organisms and their environment can affect conditions from local (e.g., a beaver pond) to global scales, over time and even after death, such as decaying logs or silica skeleton deposits from marine organisms.The process and concept of ecosystem engineering is related to niche construction, but the former relates only to the physical modifications of the habitat whereas the latter also considers the evolutionary implications of physical changes to the environment and the feedback this causes on the process of natural selection. Ecosystem engineers are defined as: organisms that directly or indirectly modulate the availability of resources to other species, by causing physical state changes in biotic or abiotic materials. In so doing they modify, maintain and create habitats.

The ecosystem engineering concept has stimulated a new appreciation for the influence that organisms have on the ecosystem and evolutionary process. The term "niche construction" is more often used in reference to the under-appreciated feedback mechanisms of natural selection imparting forces on the abiotic niche. An example of natural selection through ecosystem engineering occurs in the nests of social insects, such as ants, bees, wasps, and termites. There is an emergent feedback and balancing system in the structure of the nest that regulates, maintains and defends the physiology of the entire colony. Termite mounds, for example, maintain a constant internal temperature through the design of air-conditioning chimneys. The structure of the nests themselves are subject to the forces of natural selection. Moreover, a nest can survive over successive generations, so that progeny inherit both genetic material and a legacy niche that was constructed before their time.

Biome

Biomes are larger units of organization that categorize regions of the Earth's ecosystems, mainly according to the structure and composition of vegetation. There are different methods to define the continental boundaries of biomes dominated by different functional types of vegetative communities that are limited in distribution by climate, precipitation, weather and other environmental variables. Biomes include tropical rainforest, temperate broadleaf and mixed forest, temperate deciduous forest, taiga, tundra, hot desert, and polar desert. Other researchers have recently categorized other biomes, such as the human and oceanic microbiomes. To a microbe, the human body is a habitat and a landscape. Microbiomes were discovered largely through advances in molecular genetics, which have revealed a hidden richness of microbial diversity on the planet. The oceanic microbiome plays a significant role in the ecological biogeochemistry of the planet's oceans.

Biosphere

The largest scale of ecological organization is the biosphere: the total sum of ecosystems on the planet. Ecological relationships regulate the flux of energy, nutrients, and climate all the way up to the planetary scale. For example, the dynamic history of the planetary atmosphere's CO_2 and O_2 composition has been affected by the biogenic flux of gases coming from respiration and photosynthesis, with levels fluctuating over time in relation to the ecology and evolution of plants and animals. Ecological theory has also been used to explain self-emergent regulatory phenomena at the planetary scale: for example, the

Gaia hypothesis is an example of holism applied in ecological theory. The Gaia hypothesis states that there is an emergent feedback loop generated by the metabolism of living organisms that maintains the core temperature of the Earth and atmospheric conditions within a narrow self-regulating range of tolerance.

Individual ecology

Understanding traits of individual organisms helps explain patterns and processes at other levels of organization including populations, communities, and ecosystems. Several areas of ecology of evolution that focus on such traits are life history theory, ecophysiology, metabolic theory of ecology, and Ethology. Examples of such traits include features of an organisms life cycle such as age to maturity, life span, or metabolic costs of reproduction. Other traits may be related to structure (italicize this word), such as spines of a cactus of dorsal spines of a bluegill sunfish, or behavior (italicize this word), such as courtship or pair bonding. Other traits include emergent properties that are the result at least in part of interactions with the surrounding environment such as growth rate, resource uptake rate, winter, and deciduous vs. drought deciduous trees and shrubs.

One set of characteristics relate to body size and temperature. The metabolic theory of ecology provides a predictive qualitative set of relationships between an organism's body size and temperature and metabolic processes. In general, smaller, warmer organisms have higher metabolic rates and this results in a variety of predictions regarding individual somatic growth rates, reproduction and population growth rates, population size, and resource uptake rates.

The traits of organisms are subject to change through acclimation, development, and evolution. For this reason, individuals form a shared focus for ecology and for evolutionary ecology.

Population Ecology

Population ecology studies the dynamics of species populations and how these populations interact with the wider environment. A population consists of individuals of the same species that live, interact, and migrate through the same niche and habitat.

A primary law of population ecology is the Malthusian growth model which states, "a population will grow (or decline) exponentially as long as the environment experienced by all individuals in the population remains constant. Simplified population models usually start with four variables: death, birth, immigration, and emigration.

An example of an introductory population model describes a closed population, such as on an island, where immigration and emigration does not take place. Hypotheses are evaluated with reference to a null hypothesis which states that random processes create the observed data. In these island models, the rate of population change is described by:

$$\frac{\mathrm{d}N}{\mathrm{d}T} = bN - dN = (b - d)N = rN,$$

where N is the total number of individuals in the population, b and d are the per capita rates of birth and death respectively, and r is the per capita rate of population change.

Using these modelling techniques, Malthu's population principle of growth was later transformed into a model known as the logistic equation:

$$\frac{dN}{dT} = aN\left(1 - \frac{N}{K}\right)$$

where N is the number of individuals measured as biomass density, a is the maximum per-capita rate of change, and K is the carrying capacity of the population. The formula states that the rate of change in population size (dN/dT) is equal to growth (aN) that is limited by carrying capacity ($1 - N/K$).

Population ecology builds upon these introductory models to further understand demographic processes in real study populations. Commonly used types of data include life history, fecundity, and survivorship, and these are analysed using mathematical techniques such as matrix algebra. The information is used for managing wildlife stocks and setting harvest quotas. In cases where basic models are insufficient, ecologists may adopt different kinds of statistical methods, such as the Akaike information criterion, or use models that can become mathematically complex as several competing hypotheses are simultaneously confronted with the data.

Metapopulations and migration

The concept of metapopulations was defined in 1969 as a population of populations which go extinct locally and recolonize. Metapopulation ecology is another statistical approach that is often used in conservation research. Metapopulation models simplify the landscape into patches of varying levels of quality, and metapopulations are linked by the migratory behaviours of organisms. Animal migration is set apart from other kinds of movement; because, it involves the seasonal departure and return of individuals from a habitat. Migration is also a population-level phenomenon, as with the migration routes followed by plants as they occupied northern post-glacial environments. Plant ecologists use pollen records that accumulate and stratify in wetlands to reconstruct the timing of plant migration and dispersal relative to historic and contemporary climates. These migration routes involved an expansion of the range as plant populations expanded from one area to another. There is a larger taxonomy of movement, such as commuting, foraging, territorial behaviour, stasis, and ranging. Dispersal is usually distinguished from migration; because, it involves the one way permanent movement of individuals from their birth population into another population.

In metapopulation terminology, migrating individuals are classed as emigrants (when they leave a region) or immigrants (when they enter a region), and sites are classed either as sources or sinks. A site is a generic term that refers to places where ecologists sample populations, such as ponds or defined sampling areas in a forest. Source patches are productive sites that generate a seasonal supply of juveniles that migrate to other patch locations. Sink patches are unproductive sites that only receive migrants; the population at the site will disappear unless rescued by an adjacent source patch or environmental conditions become more favourable. Metapopulation models examine patch dynamics over time to answer potential questions about spatial and demographic ecology. The ecology of metapopulations is a dynamic process of extinction and colonization. Small patches of lower quality (i.e., sinks) are maintained or rescued by a seasonal influx of new immigrants. A dynamic metapopulation structure evolves from year to year, where some patches are sinks in dry years and are sources when conditions are more favourable. Ecologists use a mixture of computer models and field studies to explain metapopulation structure.

Community ecology

Interspecific interactions such as predation are a key aspect of community ecology.

Community ecology examines how interactions among species and their environment affect the abundance, distribution and diversity of species within communities.

Johnson & Stinchcomb (2007)

Community ecology is the study of the interactions among a collections of species that inhabit the same geographic area. Community ecologists study the determinants of patterns and processes for two or more interacting species. Research in community ecology might measure species diversity in grasslands in relation to soil fertility. It might also include the analysis of predator-prey dynamics, competition among similar plant species, or mutualistic interactions between crabs and corals.

Ecosystem Ecology

These ecosystems, as we may call them, are of the most various kinds and sizes. They form one category of the multitudinous physical systems of the universe, which range from the universe as a whole down to the atom.

Tansley (1935):

Ecosystems may be habitats within biomes that form an integrated whole and a dynamically responsive system having both physical and biological complexes. Ecosystem ecology is the science of determining the fluxes of materials (e.g. carbon, phosphorus) between different pools (e.g., tree biomass, soil organic material). Ecosystem ecologist attempt to determine the underlying causes of these fluxes. Research in ecosystem ecology might measure primary production (g C/m^2) in a wetland in relation to decomposition and consumption rates (g C/m^2/y). This requires an understanding of the community connections between plants (i.e., primary producers) and the decomposers (e.g., fungi and bacteria),

A riparian forest in the White Mountains, New Hampshire (USA) is an example of ecosystem ecology

The underlying concept of ecosystem can be traced back to 1864 in the published work of George Perkins Marsh ("Man and Nature"). Within an ecosystem, organisms are linked to the physical and biological components of their environment to which they are adapted. Ecosystems are complex adaptive systems where the interaction of life processes form self-organizing patterns across different scales of time and space. Ecosystems are broadly categorized as terrestrial, freshwater, atmospheric, or marine. Differences stem from the nature of the unique physical environments that shapes the biodiversity within each. A more recent addition to ecosystem ecology are technoecosystems, which are affected by or primarily the result of human activity.

Food webs

A food web is the archetypal ecological network. Plants capture solar energy and use it to synthesize simple sugars during photosynthesis. As plants grow, they accumulate nutrients and are eaten by grazing herbivores, and the energy is transferred through a chain of organisms by consumption. The simplified linear feeding pathways that move from a basal trophic species to a top consumer is called the food chain. The larger interlocking pattern of food chains in an ecological community creates a complex food web. Food webs are a type of concept map or a heuristic device that is used to illustrate and study pathways of energy and material flows.

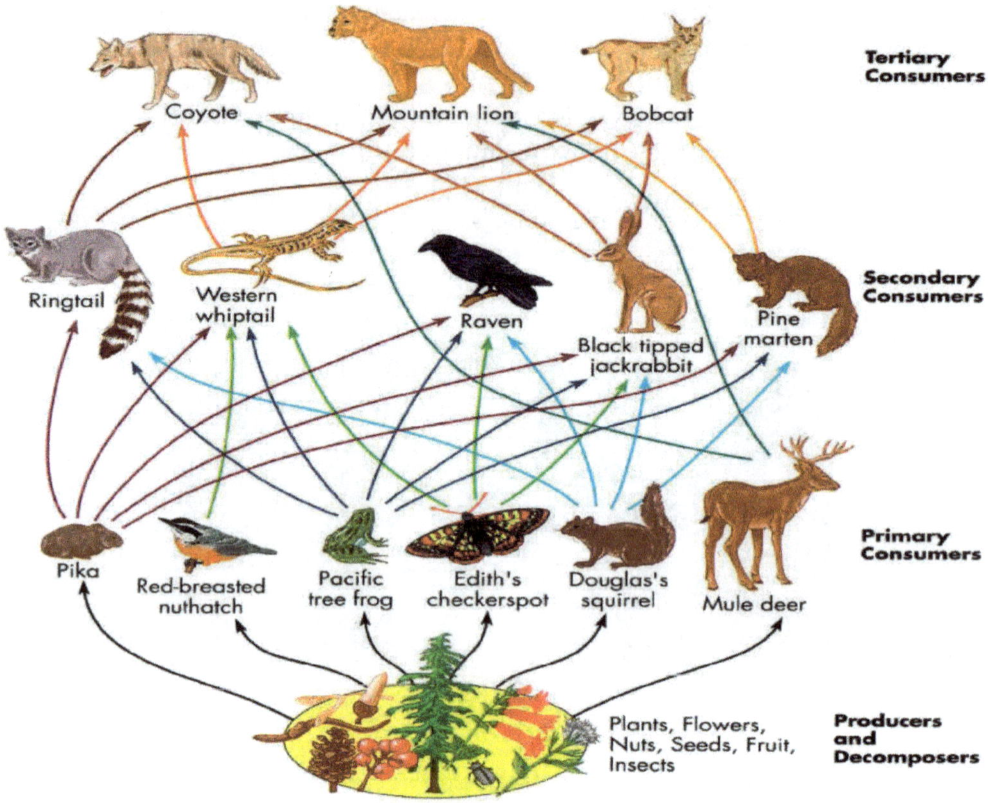

Generalized food web of waterbirds from Chesapeake Bay

Food webs are often limited relative to the real world. Complete empirical measurements are generally restricted to a specific habitat, such as a cave or a pond, and principles gleaned from food web microcosm studies are extrapolated to larger systems. Feeding relations require extensive investigations into the gut contents of organisms, which can be difficult to decipher, or stable isotopes can be used to trace the flow of nutrient diets and energy through a food web. Despite these limitations, food webs remain a valuable tool in understanding community ecosystems.

Food webs exhibit principles of ecological emergence through the nature of trophic relationships: some species have many weak feeding links (e.g., omnivores) while some are more specialized with fewer stronger feeding links (e.g., primary predators). Theoretical and empirical studies identify non-random emergent patterns of few strong and many weak linkages that explain how ecological communities remain stable over time. Food webs are composed of subgroups where members in a community are linked by strong interactions, and the weak interactions occur between these subgroups. This increases food web stability. Step by step lines or relations are drawn until a web of life is illustrated.

Trophic Levels

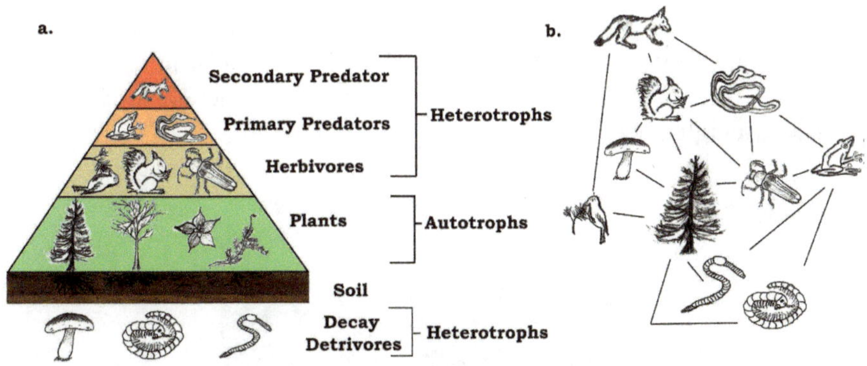

A trophic pyramid (a) and a food-web (b) illustrating ecological relationships among creatures that are typical of a northern boreal terrestrial ecosystem. The trophic pyramid roughly represents the biomass (usually measured as total dry-weight) at each level. Plants generally have the greatest biomass. Names of trophic categories are shown to the right of the pyramid. Some ecosystems, such as many wetlands, do not organize as a strict pyramid, because aquatic plants are not as productive as long-lived terrestrial plants such as trees. Ecological trophic pyramids are typically one of three kinds: 1) pyramid of numbers, 2) pyramid of biomass, or 3) pyramid of energy.

A trophic level (from Greek *troph*, τροφή, trophē, meaning "food" or "feeding") is "a group of organisms acquiring a considerable majority of its energy from the adjacent level nearer the abiotic source." Links in food webs primarily connect feeding relations or trophism among species. Biodiversity within ecosystems can be organized into trophic pyramids, in which the vertical dimension represents feeding relations that become further removed from the base of the food chain up toward top predators, and the horizontal dimension represents the abundance or biomass at each level. When the relative abundance or biomass of each species is sorted into its respective trophic level, they naturally sort into a 'pyramid of numbers'.

Species are broadly categorized as autotrophs (or primary producers), heterotrophs (or consumers), and Detritivores (or decomposers). Autotrophs are organisms that produce their own food (production is greater than respiration) by photosynthesis or chemosynthesis. Heterotrophs are organisms that must feed on others for nourishment and energy (respiration exceeds production). Heterotrophs can be further sub-divided into different functional groups, including primary consumers (strict herbivores), secondary consumers (carnivorous predators that feed exclusively on herbivores), and tertiary consumers (predators that feed on a mix of herbivores and predators). Omnivores do not fit neatly into a functional category because they eat both plant and animal tissues. It has been suggested that omnivores have a greater functional influence as predators, because compared to herbivores, they are relatively inefficient at grazing.

Trophic levels are part of the holistic or complex systems view of ecosystems. Each trophic level contains unrelated species that are grouped together because they share common ecological functions, giving a macroscopic view of the system. While the notion of trophic levels provides insight into energy flow and top-down control within food webs, it is troubled by the prevalence of omnivory in real ecosystems. This has led some ecologists to reiterate that the notion that species clearly aggregate into discrete, homogeneous trophic levels is fiction. Nonetheless, recent studies have shown that real trophic levels do exist, but above the herbivore trophic level, food webs are better characterized as a tangled web of omnivores.

Keystone Species

Sea otters, an example of a keystone species

A keystone species is a species that is connected to a disproportionately large number of other species in the food-web. Keystone species have lower levels of biomass in the trophic pyramid relative to the importance of their role. The many connections that a keystone species holds means that it maintains the organization and structure of entire communities. The loss of a keystone species results in a range of dramatic cascading effects that alters trophic dynamics, other food web connections, and can cause the extinction of other species.

Sea otters (*Enhydra lutris*) are commonly cited as an example of a keystone species; because, they limit the density of sea urchins that feed on kelp. If sea otters are removed from the system, the urchins graze until the kelp beds disappear, and this has a dramatic effect on community structure. Hunting of sea otters, for example, is thought to have led indirectly to the extinction of the Stellers sea cow (*Hydrodamalis gigas*). While the keystone species concept has been used extensively as a conservation tool, it has been criticized for being poorly defined from an operational stance. It is difficult to experimentally determine what species may hold a keystone role in each ecosystem. Furthermore, food web theory suggests that keystone species may not be common, so it is unclear how generally the keystone species model can be applied.

ECOLOGICAL COMPLEXITY

Complexity is understood as a large computational effort needed to piece together numerous interacting parts exceeding the iterative memory capacity of the human mind. Global patterns of biological diversity are complex. This biocomplexity stems from the interplay among ecological processes that operate and influence patterns at different scales that grade into each other, such as transitional areas or ecotones spanning landscapes.

Complexity stems from the interplay among levels of biological organization as energy, and matter is integrated into larger units that superimpose onto the smaller parts. "What were wholes on one level become parts on a higher one." Small scale patterns do not necessarily explain large scale phenomena, otherwise captured in the expression (coined by Aristotle) 'the sum is greater than the parts'.

Complexity in ecology is of at least six distinct types: spatial, temporal, structural, process, behavioral, and geometric. From these principles, ecologists have identified emergent and self-organizing phenomena that operate at different environmental scales of influence, ranging from molecular to planetary, and these require different explanations at each integrative level. Ecological complexity relates to the dynamic resilience of ecosystems that transition to multiple shifting steady-states directed by random fluctuations of history. Long-term ecological studies provide important track records to better understand the complexity and resilience of ecosystems over longer temporal and broader spatial scales. These studies are managed by the International Long Term Ecological Network (LTER). The longest experiment in existence is the Park Grass Experiment, which was initiated in 1856. Another example is the Hubbard Brook study, which has been in operation since 1960.

Holism remains a critical part of the theoretical foundation in contemporary ecological studies. Holism addresses the biological organization of life that self-organizes into layers of emergent whole systems that function according to non-reducible properties. This means that higher order patterns of a whole functional system, such as an ecosystem, cannot be predicted or understood by a simple summation of the parts. New properties emerge because the components interact, not because the basic nature of the components is changed.

Ecological studies are necessarily holistic as opposed to reductionistic. Holism has three scientific meanings or uses that identify with ecology: 1) the mechanistic complexity of ecosystems, 2) the practical description of patterns in quantitative reductionist terms where correlations may be identified but nothing is understood about the causal relations without reference to the whole system, which leads to 3) a metaphysical hierarchy whereby the causal relations of larger systems are understood without reference to the smaller parts. Scientific holism differs from mysticism that has appropriated the same term. An example of metaphysical holism is identified in the trend of increased exterior thickness in shells of different species. The reason for a thickness increase can be understood through reference to principles of natural selection via predation without need to reference or understand the biomolecular properties of the exterior shells.

RELATION TO EVOLUTION

Ecology and evolution are considered sister disciplines of the life sciences. Natural selection, life history, development, adaptation, populations, and inheritance are examples of concepts that thread equally into ecological and evolutionary theory. Morphological, behavioural, and genetic traits, for example, can be mapped onto evolutionary trees to study the historical development of a species in relation to their functions and roles in different ecological circumstances. In this framework, the analytical tools of ecologists and evolutionists overlap as they organize, classify, and investigate life through common systematic principals, such as phylogenetics or the Linnaean system of taxonomy. The two disciplines often appear together, such as in the title of the journal *Trends in Ecology and Evolution*. There is no sharp boundary separating ecology from evolution, and they differ more in their areas of applied focus. Both disciplines discover and explain emergent and unique properties and processes operating across different spatial or temporal scales of organization. While the boundary between ecology and evolution is not always clear, ecologists study the abiotic and biotic factors that influence evolutionary processes, and evolution can be rapid, occurring on ecological timescales as short as one generation.

Behavioural Ecology

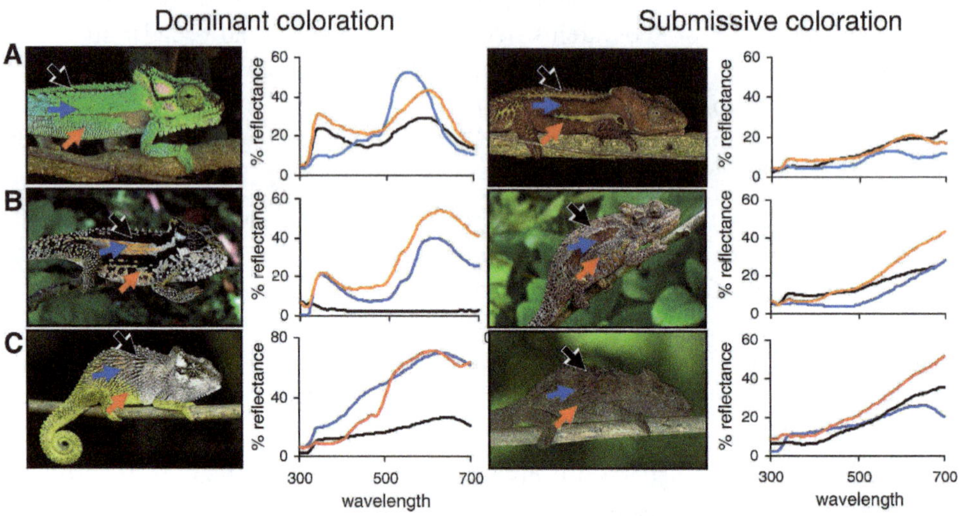

Social display and colour variation in differently adapted species of chameleons (*Bradypodion* spp.). Chameleons change their skin colour to match their background as a behavioural defence mechanism and also use colour to communicate with other members of their species, such as dominant (left) versus submissive (right) patterns shown in the three species (A-C) above.

All organisms can exhibit behaviours. Even plants express complex behaviour, including memory and communication. Behavioural ecology is the study of an organisms behaviour in its environment and its ecological and evolutionary implications. Ethology is the study of observable movement or behaviour in animals. This could include investigations of motile sperm of plants, mobile phytoplankton, zooplankton swimming toward the female egg, the cultivation of fungi by weevils, the mating dance of a salamander, or social gatherings of amoeba.

Adaptation is the central unifying concept in behavioural ecology.Behaviours can be recorded as traits and inherited in much the same way that eye and hair colour can. Behaviours can evolve by means of natural selection as adaptive traits conferring functional utilities that increases reproductive fitness.

Predator-prey interactions are an introductory concept into food-web studies as well as behavioural ecology. Prey species can exhibit different kinds of behavioural adaptations to predators, such as avoid, flee, or defend. Many prey species are faced with multiple predators that differ in the degree of danger posed. To be adapted to their environment and face predatory threats, organisms must balance their energy budgets as they invest in different aspects of their life history, such as growth, feeding, mating, socializing, or modifying their habitat. Hypotheses posited in behavioural ecology are generally based on adaptive principles of conservation, optimization, or efficiency. For example, [t]he threat-sensitive predator avoidance hypothesis predicts that prey should assess the degree of threat posed by different predators and match their behaviour according to current levels of risk or [t]he optimal flight initiation distance occurs where expected postencounter fitness is maximized, which depends on the prey's initial fitness, benefits obtainable by not fleeing, energetic escape costs, and expected fitness loss due to predation risk."

Symbiosis: Leafhoppers (*Eurymela fenestrata*) are protected by ants (*Iridomyrmex purpureus*) in a symbiotic relationship. The ants protect the leafhoppers from predators and in return the leafhoppers feeding on plants exude honeydew from their anus that provides energy and nutrients to tending ants.

Elaborate sexual displays and posturing are encountered in the behavioural ecology of animals. The birds-of-paradise, for example, sing and display elaborate ornaments during courtship. These displays serve a dual purpose of signalling healthy or well-adapted individuals and desirable genes. The displays are driven by sexual selection as an advertisement of quality of traits among suitors.

Cognitive Ecology

Cognitive ecology integrates theory and observations from evolutionary ecology and neurobiology, primarily cognitive science, in order to understand the effect that animal interaction with their habitat has on their cognitive systems and how those systems restrict behavior within an ecological and evolutionary framework. Until recently, however, cognitive scientists have not paid sufficient attention to the fundamental fact that cognitive traits evolved under particular natural settings. With consideration of the selection pressure on cognition, cognitive ecology can contribute intellectual coherence to the multidisciplinary study of cognition. As a study involving the interactions between organism and environment, cognitive ecology is closely related to enactivism, a field based upon the view that the organism and environment as bound together in reciprocal specification and selection.

Social Ecology

Social ecological behaviours are notable in social insects, slime moulds, social spiders, human society, and naked mole-rats where eusocialism has evolved. Social behaviours include reciprocally beneficial behaviours among kin and nest mates and evolve from kin and group selection. Kin selection explains altruism through genetic relationships, whereby an altruistic behaviour leading to death is rewarded by the survival of genetic copies distributed among surviving relatives. Social insects, including ants, bees, and wasps are most famously studied for this type of relationship because the male drones are clones that share the same genetic make-up as every other male in the colony. In contrast, group selectionists find examples of altruism among non-genetic relatives and explain this through selection acting on the group; whereby, it becomes selectively advantageous for groups if their members express altruistic behaviours to one another. Groups with predominantly altruistic members outdo groups with predominantly selfish members.

Coevolution

Bumblebees and the flowers they pollinate have coevolved so that both have become dependent on each other for survival.

Ecological interactions can be classified broadly into a host and an associate relationship. A host is any entity that harbours another that is called the associate. Relationships within a species that are mutually or reciprocally beneficial are called mutualisms. Examples of mutualism include fungus-growing ants employing agricultural symbiosis, bacteria living in the guts of insects and other organisms, the fig wasp and yucca moth pollination complex, lichens with fungi and photosynthetic algae, and corals with photosynthetic algae. If there is a physical connection between host and associate, the relationship is called symbiosis. Approximately 60% of all plants, for example, have a symbiotic relationship with arbuscular mycorrhizal fungi living in their roots forming an exchange network of carbohydrates for mineral nutrients.

Indirect mutualisms occur where the organisms live apart. For example, trees living in the equatorial regions of the planet supply oxygen into the atmosphere that sustains species living in distant polar regions of the planet. This relationship is called commensalism; Many others receive the benefits of clean air at no cost or harm to trees supplying the oxygen. If the associate benefits while the host suffers, the relationship is called parasitism. Although parasites impose a cost to their host (e.g., via damage to their reproductive organs or propagules, denying the services of a beneficial partner), their net effect on host fitness is not necessarily negative and, thus, becomes difficult to forecast. Co-evolution is also driven by competition among species or among members of the same species under the banner of reciprocal antagonism, such as grasses competing for growth space. The Red Queen Hypothesis, for example, posits that parasites track down and specialize on the locally common genetic defense systems of its host that drives the evolution of sexual reproduction to diversify the genetic constituency of populations responding to the antagonistic pressure.

A harvestman arachnid being parasitized by mites. The harvestman is being consumed, while the mites benefit from traveling on and feeding off of their host.

Biogeography

Biogeography (an amalgamation of *biology* and *geography*) is the comparative study of the geographic distribution of organisms and the corresponding evolution of their traits in space and time. Biogeography and ecology share many of their disciplinary roots. For example, the theory of island biogeography, published by the mathematician Robert MacArthur and ecologist Edward O. Wilson in 1967 is considered one of the fundamentals of ecological theory.

Biogeography has a long history in the natural sciences concerning the spatial distribution of plants and animals. Ecology and evolution provide the explanatory context for biogeographical studies. Biogeographical patterns result from ecological processes that influence range distributions, such as migration and dispersal. and from historical processes that split populations or species into different areas. The biogeographic processes that result in the natural splitting of species explains much of the modern distribution of the Earths biota. The splitting of lineages in a species is called vicariance biogeography and it is a sub-discipline of biogeography. There are also practical applications in the field of biogeography concerning ecological systems and processes. For example, the range and distribution of biodiversity and invasive species responding to climate change is a serious concern and active area of research in the context of global warming.

r/K-Selection theory

A population ecology concept is r/K selection theory, one of the first predictive models in ecology used to explain life-history evolution. The premise behind the r/K selection model is that natural selection pressures change according to population density. For example, when an island is first colonized, density of individuals is low. The initial increase in population size is not limited by competition, leaving an abundance of available resources for rapid population growth. These early phases of population growth experience *density-independent* forces of natural selection, which is called *r*-selection. As the population becomes more crowded, it approaches the island's carrying capacity, thus forcing individuals to compete more heavily for fewer available resources. Under crowded conditions, the population experiences density-dependent forces of natural selection, called *K*-selection.

In the *r/K*-selection model, the first variable *r* is the intrinsic rate of natural increase in population size and the second variable *K* is the carrying capacity of a population. Different species evolve different life-history strategies spanning a continuum between these two selective forces. An *r*-selected species is one that has high birth rates, low levels of parental investment, and high rates of mortality before individuals reach maturity. Evolution favours high rates of fecundity in *r*-selected species. Many kinds of insects and invasive species exhibit *r*-selected characteristics. In contrast, a *K*-selected species has low rates of fecundity, high levels of parental investment in the young, and low rates of mortality as individuals mature. Humans and elephants are examples of species exhibiting *K*-selected characteristics, including longevity and efficiency in the conversion of more resources into fewer offspring.

Molecular ecology

The important relationship between ecology and genetic inheritance predates modern techniques for molecular analysis. Molecular ecological research became more feasible with the development of rapid and accessible genetic technologies, such as the polymerase chain reaction (PCR). The rise of molecular technologies and influx of research questions into this new ecological field resulted in the publication *Molecular Ecology* in 1992. Molecular ecology uses various analytical techniques to study genes in an evolutionary and ecological context. In 1994, John Avise also played a leading role in this area of science with the publication of his book, *Molecular Markers, Natural History and Evolution*. Newer technologies opened a wave of genetic analysis into organisms once difficult to study from an ecological or evolutionary standpoint, such as bacteria, fungi, and nematodes. Molecular ecology engendered a new research paradigm for investigating ecological questions considered otherwise intractable. Molecular investigations revealed previously obscured details in the tiny intricacies of nature and improved resolution into probing questions about behavioural and biogeographical ecology. For example, molecular ecology revealed promiscuous sexual behaviour and multiple male partners in tree swallows previously thought to be socially monogamous. In a biogeographical context, the marriage between genetics, ecology, and evolution resulted in a new sub-discipline called phylogeography.

HUMAN ECOLOGY

The history of life on Earth has been a history of interaction between living things and their surroundings. To a large extent, the physical form and the habits of the earth's vegetation and its animal life have been molded by the environment. Considering the whole span of earthly time, the opposite effect, in which life actually modifies its surroundings, has been relatively slight. Only within the moment of time represented by the present century has one species man acquired significant power to alter the nature of his world.

Ecology is as much a biological science as it is a human science. Human ecology is an interdisciplinary investigation into the ecology of our species. Human ecology may be defined: (1) from a bio-ecological standpoint as the study of man as the ecological dominant in plant and animal communities and systems; (2) from a bio-ecological standpoint as simply another animal affecting and being affected by his physical environment; and (3) as a human being, somehow different from animal life in general, interacting with physical and modified environments in a distinctive and creative way. A truly interdisciplinary human ecology will most likely address itself to all three. The term was formally introduced in 1921, but many sociologists, geographers, psychologists, and other disciplines were interested in human relations to natural systems centuries prior, especially in the late 19th century.

The ecological complexities human beings are facing through the technological transformation of the planetary biome has brought on the Anthropocene. The unique set of circumstances has generated the need for a new unifying science called coupled human and natural systems that builds upon, but moves beyond the field of human ecology. Ecosystems tie into human societies through the critical and all encompassing life-supporting functions they sustain. In recognition of these functions and the incapability of traditional economic valuation methods to see the value in ecosystems, there has been a surge of interest in social-natural capital, which provides the means to put a value on the stock and use of information and materials stemming from ecosystem goods and services. Ecosystems produce, regulate, maintain, and supply services of critical necessity to human wellbeing and economic needs of societies. Ecosystems relate importantly to human ecology as they are the ultimate base foundation of global economics as every commodity and capacity for exchange ultimately stems from the ecosystems on Earth.

Restoration and management

Ecosystem management is not just about science nor is it simply an extension of traditional resource management; it offers a fundamental reframing of how humans may work with nature.

Ecology is an employed science of restoration, repairing disturbed sites through human intervention, in natural resource management, and in environmental impact assessments. Edward O. Wilson predicted in 1992 that the 21st century "will be the era of restoration in ecology". Ecological science has boomed in the industrial investment of restoring ecosystems and their processes in abandoned sites after disturbance. Natural resource managers, in forestry, for example, employ ecologists to develop, adapt, and implement ecosystem based methods into the planning, operation, and restoration phases of land-use. Ecological science is used in the methods of sustainable harvesting, disease, and fire outbreak management, in fisheries stock management, for integrating land-use with protected areas and communities, and conservation in complex geo-political landscapes.

RELATION TO THE ENVIRONMENT

The environment of ecosystems includes both physical parameters and biotic attributes. It is dynamically interlinked, and contains resources for organisms at any time throughout their life cycle. Like "ecology", the term "environment" has different conceptual meanings and overlaps with the concept of "nature". Environment includes the physical world, the social world of human relations and the built world of human creation. The physical environment is external to the level of biological organization under investigation, including abiotic factors such as temperature, radiation, light, chemistry, climate and geology. The biotic environment includes genes, cells, organisms, members of the same species (conspecifics) and other species that share a habitat.

The distinction between external and internal environments, however, is an abstraction parsing life and environment into units or facts that are inseparable in reality. There is an interpenetration of cause and effect between the environment and life. The laws of thermodynamics, for example, apply to ecology by means of its physical state. With an understanding of metabolic and thermodynamic principles, a complete accounting of energy and material flow can be traced through an ecosystem. In this way, the environmental and ecological relations are studied through reference to conceptually manageable and isolated material parts. After the effective environmental components are understood through reference to their causes; however, they conceptually link back together as an integrated whole, or *holocoenotic* system as it was once called. This is known as the dialectical approach to ecology. The dialectical approach examines the parts, but integrates the organism and the environment into a dynamic whole (or umwelt). Change in one ecological or environmental factor can concurrently affect the dynamic state of an entire ecosystem.

Disturbance and Resilience

Ecosystems are regularly confronted with natural environmental variations and disturbances over time and geographic space. A disturbance is any process that removes biomass from a community, such as a fire, flood, drought, or predation. Disturbances occur over vastly different ranges in terms of magnitudes as well as distances and time periods, and are both the cause and product of natural fluctuations in death rates, species assemblages, and biomass densities within an ecological community. These disturbances create places of renewal where new directions emerge from the patchwork of natural experimentation and opportunity. Ecological resilience is a cornerstone theory in ecosystem management. Biodiversity fuels the resilience of ecosystems acting as a kind of regenerative insurance.

The Earth's Early Atmosphere

The Earth was formed approximately 4.5 billion years ago. As it cooled and a crust and oceans formed, its atmosphere transformed from being dominated by hydrogen to one composed mostly of methane and ammonia. Over the next billion years, the metabolic activity of life transformed the atmosphere into a mixture of carbon dioxide, nitrogen, and water vapor. These gases changed the way that light from the sun hit the Earths surface and greenhouse effects trapped heat. There were untapped sources of free energy within the mixture of reducing and oxidizing gasses that set the stage for primitive ecosystems to evolve and, in turn, the atmosphere also evolved.

The leaf is the primary site of photosynthesis in most plants.

Throughout history, the Earths atmosphere and biogeochemical cycles have been in a dynamic equilibrium with planetary ecosystems. The history is characterized by periods of significant transformation followed by millions of years of stability. The evolution of the earliest organisms, likely anaerobic methanogen microbes, started the process by converting atmospheric hydrogen into methane ($4H_2 + CO_2 \rightarrow CH_4 + 2H_2O$). Anoxygenic photosynthesis reduced hydrogen concentrations and increased atmospheric methane, by converting hydrogen sulfide into water or other sulfur compounds (for example, $2H_2S + CO_2 + h\nu \rightarrow CH_2O + H_2O + 2S$). Early forms of fermentation also increased levels of atmospheric methane. The transition to an oxygen-dominant atmosphere (the *Great Oxidation*) did not begin until approximately 2.4–2.3 billion years ago, but photosynthetic processes started 0.3 to 1 billion years prior.

Radiation: Heat, Temperature and Light

The biology of life operates within a certain range of temperatures. Heat is a form of energy that regulates temperature. Heat affects growth rates, activity, behaviour, and primary production. Temperature is largely dependent on the incidence of solar radiation. The latitudinal and longitudinal spatial variation of temperature greatly affects climates and consequently the distribution of biodiversity and levels of primary production in different ecosystems or biomes across the planet. Heat and temperature relate importantly to metabolic activity. Poikilotherms, for example, have a body temperature that is largely regulated and dependent on the temperature of the external environment. In contrast, homeotherms regulate their internal body temperature by expending metabolic energy.

There is a relationship between light, primary production, and ecological energy budgets. Sunlight is the primary input of energy into the planets ecosystems. Light is composed of electromagnetic energy of different wavelengths. Radiant energy from the sun generates heat, provides photons of light measured as active energy in the chemical reactions of life, and also acts as a catalyst for genetic mutation. Plants, algae, and some bacteria absorb light and assimilate the energy through photosynthesis. Organisms capable of assimilating energy by photosynthesis or through inorganic fixation of H_2S are autotrophs. Autotrophs — responsible for primary production — assimilate light energy which becomes metabolically stored as potential energy in the form of biochemical enthalpic bonds.

Physical Environments

Water

Diffusion of carbon dioxide and oxygen is approximately 10,000 times slower in water than in air. When soils are flooded, they quickly lose oxygen, becoming hypoxic (an environment with O_2 concentration below 2 mg/liter) and eventually completely anoxic where anaerobic bacteria thrive among the roots. Water also influences the intensity and spectral composition of light as it reflects off the water surface and submerged particles. Aquatic plants exhibit a wide variety of morphological and physiological adaptations that allow them to survive, compete, and diversify in these environments. For example, their roots and stems contain large air spaces (aerenchyma) that regulate the efficient transportation of gases (for example, CO_2 and O_2) used in respiration and photosynthesis. Salt water plants (halophytes) have additional specialized adaptations, such as the development of special organs for shedding salt and osmoregulating their internal salt (NaCl) concentrations, to live in estuarine, brackish, or oceanic environments. Anaerobic soil microorganisms in aquatic environments use nitrate, manganese ions, ferric ions, sulfate, carbon dioxide, and some organic compounds; other microorganisms are facultative anaerobes and use oxygen during respiration when the soil becomes drier. The activity of soil microorganisms and the chemistry of the water reduces the oxidation-reduction potentials of the water. Carbon dioxide, for example, is reduced to methane (CH_4) by methanogenic bacteria. The physiology of fish is also specially adapted to compensate for environmental salt levels through osmoregulation. Their gills form electrochemical gradients that mediate salt excretion in salt water and uptake in fresh water.

Gravity

The shape and energy of the land is significantly affected by gravitational forces. On a large scale, the distribution of gravitational forces on the earth is uneven and influences the shape and movement of tectonic plates as well as influencing geomorphic processes such as orogeny and erosion. These forces govern many of the geophysical properties and distributions of ecological biomes across the Earth. On the organismal scale, gravitational forces provide directional cues for plant and fungal growth (gravitropism), orientation cues for animal migrations, and influence the biomechanics and size of animals. Ecological traits, such as allocation of biomass in trees during growth are subject to mechanical failure as gravitational forces influence the position and structure of branches and leaves. The cardiovascular systems of animals are functionally adapted to overcome pressure and gravitational forces that change according to the features of organisms (e.g., height, size, shape), their behaviour (e.g., diving, running, flying), and the habitat occupied (e.g., water, hot deserts, cold tundra).

Pressure

Climatic and osmotic pressure places physiological constraints on organisms, especially those that fly and respire at high altitudes, or dive to deep ocean depths. These constraints influence vertical limits of ecosystems in the biosphere, as organisms are physiologically sensitive and adapted to atmospheric and osmotic water pressure differences. For example, oxygen levels decrease with decreasing pressure and are a limiting factor for life at higher altitudes. Water transportation by plants is another important ecophysiological process affected by osmotic pressure gradients. Water pressure in the depths of oceans requires that organisms adapt to these conditions. For example, diving animals such as whales, dolphins, and seals are specially adapted to deal with changes in sound due to water pressure differences. Differences between hagfish species provide another example of adaptation to deep-sea pressure through specialized protein adaptations.

Wind and Turbulence

The architecture of the inflorescence in grasses is subject to the physical pressures of wind and shaped by the forces of natural selection facilitating wind-pollination (anemophily).

Turbulent forces in air and water affect the environment and ecosystem distribution, form and dynamics. On a planetary scale, ecosystems are affected by circulation patterns in the global trade winds. Wind power and the turbulent forces it creates can influence heat, nutrient, and biochemical profiles of ecosystems. For example, wind running over the surface of a lake creates turbulence, mixing the water column and influencing the environmental profile to create thermally layered zones, affecting how fish, algae, and other parts of the aquatic ecosystem are structured. Wind speed and turbulence also influence evapotranspiration rates and energy budgets in plants and animals. Wind speed, temperature and moisture content can vary as winds travel across different land features and elevations. For example, the westerlies come into contact with the coastal and interior mountains of western North America to produce a rain shadow on the leeward side of the mountain. The air expands and moisture condenses as the winds increase in elevation; this is called orographic lift and can cause precipitation. This environmental process produces spatial divisions in biodiversity, as species adapted to wetter conditions are range-restricted to the coastal mountain valleys and unable to migrate across the xeric ecosystems (e.g., of the Columbia Basin in western North America) to intermix with sister lineages that are segregated to the interior mountain systems.

Fire

Forest fires modify the land by leaving behind an environmental mosaic that diversifies the landscape into different seral stages and habitats of varied quality (left). Some species are adapted to forest fires, such as pine trees that open their cones only after fire exposure (right).

Plants convert carbon dioxide into biomass and emit oxygen into the atmosphere. By approximately 350 million years ago (the end of the Devonian period), photosynthesis had brought the concentration of atmospheric oxygen above 17%, which allowed combustion to occur. Fire releases CO_2 and converts fuel into ash and tar. Fire is a significant ecological parameter that raises many issues pertaining to its control and suppression. While the issue of fire in relation to ecology and plants has been recognized for a long time, Charles Cooper brought attention to the issue of forest fires in relation to the ecology of forest fire suppression and management in the 1960s.

Native North Americans were among the first to influence fire regimes by controlling their spread near their homes or by lighting fires to stimulate the production of herbaceous foods and basketry materials. Fire creates a heterogeneous ecosystem age and canopy structure, and the altered soil nutrient supply and cleared canopy structure opens new ecological niches for seedling establishment. Most ecosystems are adapted to natural fire cycles. Plants, for example, are equipped with a variety of adaptations to deal with forest fires. Some species (e.g., *Pinus halepensis*) cannot germinate until after their seeds have lived through a fire or been exposed to certain compounds from smoke. Environmentally triggered germination of seeds is called serotiny. Fire plays a major role in the persistence and resilience of ecosystems.

Soils

Soil is the living top layer of mineral and organic dirt that covers the surface of the planet. It is the chief organizing centre of most ecosystem functions, and it is of critical importance in agricultural science and ecology. The decomposition of dead organic matter (for example, leaves on the forest floor), results in soils containing minerals and nutrients that feed into plant production. The whole of the planet's soil ecosystems is called the pedosphere where a large biomass of the Earth's biodiversity organizes into trophic levels. Invertebrates that feed and shred larger leaves, for example, create smaller bits for smaller organisms in the feeding chain. Collectively, these organisms are the detritivores that regulate soil formation. Tree roots, fungi, bacteria, worms, ants, beetles, centipedes, spiders, mammals, birds, reptiles, amphibians, and other less familiar creatures all work to create the trophic web of life in soil ecosystems. Soils form composite phenotypes where inorganic matter is enveloped into the physiology of a whole community. As organisms feed and migrate through soils they physically displace materials, an ecological process called bioturbation. This aerates soils and stimulates heterotrophic growth and production. Soil microorganisms are influenced by and feed back into the trophic dynamics of the ecosystem. No single axis of causality can be discerned to segregate the biological from geomorphological systems in soils. Paleoecological studies of soils places the origin for bioturbation to a time before the Cambrian period. Other events, such as the evolution of trees and the colonization of land in the Devonian period played a significant role in the early development of ecological trophism in soils.

Biogeochemistry and climate

Ecologists study and measure nutrient budgets to understand how these materials are regulated, flow, and recycled through the environment. This research has led to an understanding that there is global feedback between ecosystems and the physical parameters of this planet, including minerals, soil, pH, ions, water, and atmospheric gases. Six major elements (hydrogen, carbon, nitrogen, oxygen, sulfur, and phosphorus; H, C, N, O, S, and P) form the constitution of all biological macromolecules and feed into the Earths geochemical processes. From the smallest scale of biology, the combined effect of billions upon billions of ecological processes amplify and ultimately regulate the biogeochemical cycles of the Earth. Understanding the relations and cycles mediated between these elements and their ecological pathways has significant bearing toward understanding global biogeochemistry.

The ecology of global carbon budgets gives one example of the linkage between biodiversity and biogeochemistry. It is estimated that the Earths oceans hold 40,000 gigatonnes (Gt) of carbon, that vegetation and soil hold 2070 Gt, and that fossil fuel emissions are 6.3 Gt carbon per year. There have been major restructurings in these global carbon budgets during the Earths history, regulated to a large extent by the ecology of the land. For example, through the early-mid Eocene volcanic outgassing, the oxidation of methane stored in wetlands, and seafloor gases increased atmospheric CO_2 (carbon dioxide) concentrations to levels as high as 3500 ppm.

In the Oligocene, from twenty-five to thirty-two million years ago, there was another significant restructuring of the global carbon cycle as grasses evolved a new mechanism of photosynthesis, C_4 photosynthesis, and expanded their ranges. This new pathway evolved in response to the drop in atmospheric CO_2 concentrations below 550 ppm. The relative abundance and distribution of biodiversity alters the dynamics between organisms and their environment such that ecosystems can be both cause and effect in relation to climate change. Human-driven modifications to the planet's ecosystems (e.g., disturbance, biodiversity loss, agriculture) contributes to rising atmospheric greenhouse gas levels. Transformation of the global carbon cycle in the next century is projected to raise planetary temperatures, lead to more extreme fluctuations in weather, alter species distributions, and increase extinction rates. The effect of global warming is already being registered in melting glaciers, melting mountain ice caps, and rising sea levels. Consequently, species distributions are changing along waterfronts and in continental areas where migration patterns and breeding grounds are tracking the prevailing shifts in climate. Large sections of permafrost are also melting to create a new mosaic of flooded areas having increased rates of soil decomposition activity that raises methane (CH_4)

emissions. There is concern over increases in atmospheric methane in the context of the global carbon cycle, because methane is a greenhouse gas that is 23 times more effective at absorbing long-wave radiation than CO_2 on a 100-year time scale. Hence, there is a relationship between global warming, decomposition and respiration in soils and wetlands producing significant climate feedbacks and globally altered biogeochemical cycles.

HISTORY

Early Beginnings

Ecology has a complex origin, due in large part to its interdisciplinary nature. Ancient Greek philosophers such as Hippocrates and Aristotle were among the first to record observations on natural history. However, they viewed life in terms of essentialism, where species were conceptualized as static unchanging things while varieties were seen as aberrations of an idealized type. This contrasts against the modern understanding of ecological theory where varieties are viewed as the real phenomena of interest and having a role in the origins of adaptations by means of natural selection. Early conceptions of ecology, such as a balance and regulation in nature can be traced to Herodotus (died *c.* 425 BC), who described one of the earliest accounts of mutualism in his observation of "natural dentistry". Basking Nile crocodiles, he noted, would open their mouths to give sandpipers safe access to pluck leeches out, giving nutrition to the sandpiper and oral hygiene for the crocodile. Aristotle was an early influence on the philosophical development of ecology. He and his student Theophrastus made extensive observations on plant and animal migrations, biogeography, physiology, and on their behaviour, giving an early analogue to the modern concept of an ecological niche.

Ecological concepts such as food chains, population regulation, and productivity were first developed in the 1700s, through the published works of microscopist Antoni van Leeuwenhoek (1632–1723) and botanist Richard Bradley (1688?–1732). Biogeographer Alexander von Humboldt (1769–1859) was an early pioneer in ecological thinking and was among the first to recognize ecological gradients, where species are replaced or altered in form along environmental gradients, such as a cline forming along a rise in elevation. Humboldt drew inspiration from Isaac Newton as he developed a form of "terrestrial physics". In Newtonian fashion, he brought a scientific exactitude for measurement into natural history and even alluded to concepts that are the foundation of a modern ecological law on species-to-area relationships. Natural historians, such as Humboldt, James Hutton, and Jean-Baptiste Lamarck (among others) laid the foundations of the modern ecological sciences. The term "ecology" (German: *Oekologie, Ökologie*) is of a more recent origin and was first coined by the German biologist Ernst Haeckel in his book *Generelle Morphologie der Organismen* (1866). Haeckel was a zoologist, artist, writer, and later in life a professor of comparative anatomy.

By ecology, we mean the whole science of the relations of the organism to the environment including, in the broad sense, all the conditions of existence. Thus the theory of evolution explains the housekeeping relations of organisms mechanistically as the necessary consequences of effectual causes and so forms the monistic groundwork of ecology.

Ernst Haeckel (left) and Eugenius Warming (right), two founders of ecology

Opinions differ on who was the founder of modern ecological theory. Some mark Haeckels definition as the beginning; others say it was Eugenius Warming with the writing of Oecology of Plants: An Introduction to the Study of Plant Communities (1895), or Carl Linnaeus principles on the economy of nature that matured in the early 18th century. Linnaeus founded an early branch of ecology that he called the economy of nature. His works influenced Charles Darwin, who adopted Linnaeus phrase on the *economy or polity of nature* in *The Origin of Species*. Linnaeus was the first to frame the balance of nature as a testable hypothesis. Haeckel, who admired Darwin's work, defined ecology in reference to the economy of nature, which has led some to question whether ecology and the economy of nature are synonymous.

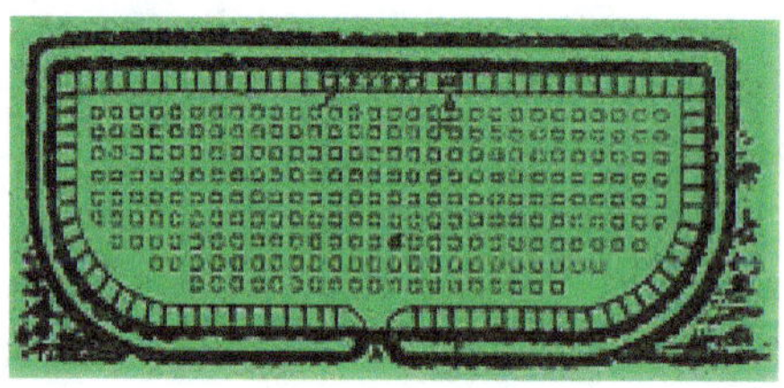

The layout of the first ecological experiment, carried out in a grass garden at Woburn Abbey in 1816, was noted by Charles Darwin in *The Origin of Species*. The experiment studied the performance of different mixtures of species planted in different kinds of soils.

From Aristotle until Darwin, the natural world was predominantly considered static and unchanging. Prior to *The Origin of Species*, there was little appreciation or understanding of the dynamic and reciprocal relations between organisms, their adaptations, and the environment. An exception is the 1789 publication *Natural History of Selborne* by Gilbert White (1720–1793), considered by some to be one of the earliest texts on ecology. While Charles Darwin is mainly noted for his treatise on evolution, he was one of the founders of soil ecology, and he made note of the first ecological experiment in *The Origin of Species*. Evolutionary theory changed the way that researchers approached the ecological sciences.

Since 1900

Modern ecology is a young science that first attracted substantial scientific attention toward the end of the 19th century (around the same time that evolutionary studies were gaining scientific interest). Notable scientist Ellen Swallow Richards may have first introduced the term "oekology (which eventually morphed into home economics) in the U.S. as early 1892.

In the early 20th century, ecology transitioned from a more descriptive form of natural history to a more analytical form of *scientific natural history*.Frederic Clements published the first American ecology book in 1905,presenting the idea of plant communities as a superorganism. This publication launched a debate between ecological holism and individualism that lasted until the 1970s. Clements superorganism concept proposed that ecosystems progress through regular and determined stages of seral development that are analogous to the developmental stages of an organism. The Clementsian paradigm was challenged by Henry Gleason, who stated that ecological communities develop from the unique and coincidental association of individual organisms. This perceptual shift placed the focus back onto the life histories of individual organisms and how this relates to the development of community associations.

The Clementsian superorganism theory was an overextended application of an idealistic form of holism. The term "holism" was coined in 1926 by Jan Christiaan Smuts, a South African general and polarizing historical figure who was inspired by Clements superorganism concept. Around the same time, Charles Elton pioneered the concept of food chains in his classical book *Animal Ecology*. Elton defined ecological relations using concepts of food chains, food cycles, and food size, and described numerical relations among different functional groups and their relative abundance. Elton's 'food cycle' was replaced by 'food web' in a subsequent ecological text. Alfred J. Lotka brought in many theoretical concepts applying thermodynamic principles to ecology.

In 1942, Raymond Lindeman wrote a landmark paper on the trophic dynamics of ecology, which was published posthumously after initially being rejected for its theoretical emphasis. Trophic dynamics became the foundation for much of the work to follow on energy and material flow through ecosystems. Robert MacArthur advanced mathematical theory, predictions, and tests in ecology in the 1950s, which inspired a resurgent school of theoretical mathematical ecologists. Ecology also has developed through contributions from other nations, including Russia's Vladimir Vernadsky and his founding of the biosphere concept in the 1920s and Japan's Kinji Imanishi and his concepts of harmony in nature and habitat segregation in the 1950s. Scientific recognition of contributions to ecology from non-English-speaking cultures is hampered by language and translation barriers.

Ecology surged in popular and scientific interest during the 1960–1970s environmental movement. There are strong historical and scientific ties between ecology, environmental management, and protection. The historical emphasis and poetic naturalistic writings advocating the protection of wild places by notable ecologists in the history of conservation biology, such as Aldo Leopold and Arthur Tansley, have been seen as far removed from urban centres where, it is claimed, the concentration of pollution and environmental degradation is located. Palamar (2008) notes an overshadowing by mainstream environmentalism of pioneering women in the early 1900s who fought for urban health ecology (then called euthenics) and brought about changes in environmental legislation. Women such as Ellen Swallow Richards and Julia Lathrop, among others, were precursors to the more popularized environmental movements after the 1950s.

In 1962, marine biologist and ecologist Rachel Carsons book *Silent Spring* helped to mobilize the environmental movement by alerting the public to toxic pesticides, such as DDT, bioaccumulating in the environment. Carson used ecological science to link the release of environmental toxins to human and ecosystem health. Since then, ecologists have worked to bridge their understanding of the degradation of the planets ecosystems with environmental politics, law, restoration, and natural resources management.

REFERENCES

1. Eric Laferrière; Peter J. Stoett (2 September 2003). International Relations Theory and Ecological Thought: Towards a Synthesis. Routledge. pp. 25–. ISBN 978-1-134-71068-3.
2. Stadler, B.; Michalzik, B.; Müller, T. (1998). "Linking aphid ecology with nutrient fluxes in a coniferous forest". Ecology. 79 (5): 1514–1525. doi:10.1890/0012-9658(1998)079[1514:LAEWNF]2.0.CO;2. ISSN 0012-9658.
3. Humphreys, N. J.; Douglas, A. E. (1997). "Partitioning of symbiotic bacteria between generations of an insect: a quantitative study of a Buchnera sp. in the pea aphid (Acyrthosiphon pisum) reared at different temperatures". Applied and Environmental Microbiology. 63 (8): 3294–3296. PMC 1389233. PMID 16535678.
4. Liere, Heidi; Jackson, Doug; Vandermeer, John; Wilby, Andrew (20 September 2012). "Ecological Complexity in a Coffee Agroecosystem: Spatial Heterogeneity, Population Persistence and Biological Control". PLoS ONE. 7 (9): e45508. Bibcode:2012PLoSO...745508L. doi:10.1371/journal.pone.0045508. PMC 3447771. PMID 23029061.
5. Odum, E. P.; Barrett, G. W. (2005). Fundamentals of Ecology. Brooks Cole. p. 598. ISBN 978-0-534-42066-6.
6. O'Neill, D. L.; Deangelis, D. L.; Waide, J. B.; Allen, T. F. H. (1986). A Hierarchical Concept of Ecosystems. Princeton University Press. p. 253. ISBN 0-691-08436-X.
7. Nachtomy, Ohad; Shavit, Ayelet; Smith, Justin (2002). "Leibnizian organisms, nested individuals, and units of selection". Theory in Biosciences. 121 (2): 205–230. doi:10.1007/s12064-002-0020-9.
8. Holling, C. S. (2004). "Understanding the complexity of economic, ecological, and social systems". Ecosystems. 4 (5): 390–405. doi:10.1007/s10021-001-0101-5.
9. Levin, S. A. (1999). Fragile Dominion: Complexity and the Commons. Reading, MA: Perseus Books. ISBN 978-0-7382-0319-5.
10. Noss, R. F.; Carpenter, A. Y. (1994). Saving Nature's Legacy: Protecting and Restoring Biodiversity. Island Press. p. 443. ISBN 978-1-55963-248-5.

11. Noss, R. F. (1990). "Indicators for monitoring biodiversity: A hierarchical approach". Conservation Biology. 4 (4): 355–364. doi:10.1111/j.1523-1739.1990.tb00309.x. JSTOR 2385928.

12. Scholes, R. J.; Mace, G. M.; Turner, W.; Geller, G. N.; Jürgens, N.; Larigauderie, A.; Muchoney, D.; Walther, B. A.; Mooney, H. A. (2008). "Toward a global biodiversity observing system" (PDF). Science. 321 (5892): 1044–1045. doi:10.1126/science.1162055. PMID 18719268.

13. Cardinale, Bradley J.; Duffy, J. Emmett; Gonzalez, Andrew; Hooper, David U.; Perrings, Charles; Venail, Patrick; Narwani, Anita; Mace, Georgina M.; Tilman, David; Wardle, David A.; Kinzig, Ann P.; Daily, Gretchen C.; Loreau, Michel; Grace, James B.; Larigauderie, Anne; Srivastava, Diane S.; Naeem, Shahid; Gonzalez, Andrew; Hooper, David U.; Perrings, Charles; Venail, Patrick; Narwani, Anita; Mace, Georgina M.; Tilman, David; Wardle, David A.; Kinzig, Ann P.; Daily, Gretchen C.; Loreau, Michel; Grace, James B.; Larigauderie, Anne; Srivastava, Diane S.; Naeem, Shahid (6 June 2012). "Biodiversity loss and its impact on humanity". Nature. 486 (7401): 59–67. Bibcode:2012Natur.486...59C. doi:10.1038/nature11148. PMID 22678280.

14. Wilson, E. O. (2000). "A global biodiversity map". Science. 289 (5488): 2279. PMID 11041790.

15. Purvis, A.; Hector, A. (2000). "Getting the measure of biodiversity" (PDF). Nature. 405 (6783): 212–218. doi:10.1038/35012221. PMID 10821281.

16. Ostfeld, R. S. (2009). "Biodiversity loss and the rise of zoonotic pathogens" (PDF). Clinical Microbiology and Infection. 15 (s1): 40–43. doi:10.1111/j.1469-0691.2008.02691.x. PMID 19220353.

17. Tierney, Geraldine L.; Faber-Langendoen, Don; Mitchell, Brian R.; Shriver, W. Gregory; Gibbs, James P. (2009). "Monitoring and evaluating the ecological integrity of forest ecosystems" (PDF). Frontiers in Ecology and the Environment. 7 (6): 308–316. doi:10.1890/070176.

18. Ceballos, G.; Ehrlich, P. R. (2002). "Mammal population losses and the extinction crisis" (PDF). Science. 296 (5569): 904–907. Bibcode:2002Sci...296..904C. doi:10.1126/science.1069349. PMID 11988573. Retrieved 2010-03-16.

19. Palumbi, Stephen R; S. R., Paul A; Allan, J. David; Beck, Michael W.; Fautin, Daphne G.; Fogarty, Michael J.; Halpern, Benjamin S.; Incze, Lewis S.; Leong, Jo-Ann; et al. (2009). "Managing for ocean biodiversity to sustain marine ecosystem services" (PDF). Frontiers in Ecology and the Environment. 7 (4): 204–211. doi:10.1890/070135.

20. Wilcove, D. S.; Wikelski, M. (2008). "Going, going, gone: Is animal migration disappearing". PLoS Biology. 6 (7): e188. doi:10.1371/journal.pbio.0060188. PMC 2486312 . PMID 18666834.

21. Hammond, H. (2009). Maintaining Whole Systems on the Earth's Crown: Ecosystem-based Conservation Planning for the Boreal Forest. Slocan Park, BC: Silva Forest Foundation. p. 380. ISBN 978-0-9734779-0-0.

22. Whittaker, R. H.; Levin, S. A.; Root, R. B. (1973). "Niche, habitat, and ecotope" (PDF). The American Naturalist. 107 (955): 321–338. doi:10.1086/282837.
23. Beyer, Hawthorne, L.; Haydon, Daniel, T.; Morales, Juan M.; Frair, Jacqueline L.; Hebblewhite, Mark; Mitchell, Michael; Matthiopoulos, Jason (2010). "The interpretation of habitat preference metrics under use–availability designs". Philosophical Transactions of the Royal Society B. 365 (1550): 2245–2254. doi:10.1098/rstb.2010.0083. PMC 2894962 . PMID 20566501.
24. Schoener, T. W. (1975). "Presence and absence of habitat shift in some widespread lizard species". Ecological Monographs. 45 (3): 233–258. doi:10.2307/1942423. JSTOR 1942423.
25. Vitt, L. J.; Caldwell, J. P.; Zani, P. A.; Titus, T. A. (1997). "The role of habitat shift in the evolution of lizard morphology: Evidence from tropical Tropidurus". Proceedings of the National Academy of Sciences. 94 (8): 3828–3832. Bibcode:1997PNAS...94.3828V. doi:10.1073/pnas.94.8.3828. PMC 20526 . PMID 9108063.
26. Kiessling, W.; Simpson, C.; Foote, M. (2009). "Reefs as cradles of evolution and sources of biodiversity in the Phanerozoic". Science. 327 (5962): 196–198. Bibcode:2010Sci...327..196K. doi:10.1126/science.1182241. PMID 20056888.
27. Laland, K. N.; Odling-Smee, F. J.; Feldman, M. W. (1999). "Evolutionary consequences of niche construction and their implications for ecology". Proceedings of the National Academy of Sciences. 96 (18): 10242–10247. Bibcode:1999PNAS...9610242L. doi:10.1073/pnas.96.18.10242. PMC 17873 . PMID 10468593.
28. Hughes, D. P.; Pierce, N. E.; Boomsma, J. J. (2008). "Social insect symbionts: evolution in homeostatic fortresses" (PDF). Trends in Ecology & Evolution. 23 (12): 672–677. doi:10.1016/j.tree.2008.07.011. PMID 18951653.
29. Wiens, J. J.; Graham, C. H. (2005). "Niche conservatism: Integrating evolution, ecology, and conservation biology" (PDF). Annual Review of Ecology, Evolution, and Systematics. 36: 519–539. doi:10.1146/annurev.ecolsys.36.102803.095431.
30. Hutchinson, G. E. (1957). A Treatise on Limnology. New York, NY: Wiley. p. 1015. ISBN 0-471-42572-9.
31. Hutchinson, G. E. (1957). "Concluding remarks" (PDF). Cold Spring Harbor Symposia on Quantitative Biology. 22: 415–427. doi:10.1101/SQB.1957.022.01.039.
32. Begon, M.; Townsend, C. R.; Harper, J. L. (2005). Ecology: From Individuals to Ecosystems (4th ed.). Wiley-Blackwell. p. 752. ISBN 1-4051-1117-8.
33. D. L., Hardesty (1975). "The niche concept: suggestions for its use in human ecology". Human Ecology. 3 (2): 71–85. doi:10.1007/BF01552263. JSTOR 4602315.
34. Pearman, P. B.; Guisan, A.; Broennimann, O.; Randin, C. F. (2008). "Niche dynamics in space and time" (PDF). Trends in Ecology & Evolution. 23 (3): 149–158. doi:10.1016/j.tree.2007.11.005. PMID 18289716.
35. Levins, R.; Lewontin, R. (1980). "Dialectics and reductionism in ecology"

(PDF). Synthese. 43: 47–78. doi:10.1007/bf00413856.

36. Hardin, G. (1960). "The competitive exclusion principal". Science. 131 (3409): 1292–1297. Bibcode:1960Sci...131.1292H. doi:10.1126/science.131.3409.1292. PMID 14399717.
37. Scheffer, M.; van Nes, E. H. (2006). "Self-organized similarity, the evolutionary emergence of groups of similar species" (PDF). Proceedings of the National Academy of Sciences. 103 (16): 6230–6235. Bibcode:2006PNAS..103.6230S. doi:10.1073/pnas.0508024103.
38. Hastings, Alan; Byers, James E.; Crooks, Jeffrey A.; Cuddington, Kim; Jones, Clive G.; Lambrinos, John G.; Talley, Theresa S.; Wilson, William G. (2007). "Ecosystem engineering in space and time". Ecology Letters. 10 (2): 153–164. doi:10.1111/j.1461-0248.2006.00997.x. PMID 17257103.
39. Jones, Clive G.; Lawton, John H.; Shachak, Moshe (1994). "Organisms as ecosystem engineers". Oikos. 69 (3): 373–386. doi:10.2307/3545850. JSTOR 3545850.
40. Wright, J. P.; Jones, C.G. (2006). "The concept of organisms as ecosystem engineers ten years on: Progress, limitations, and challenges". BioScience. 56 (3): 203–209. doi:10.1641/0006-3568(2006)056[0203:TCOOAE]2.0.CO;2. ISSN 0006-3568.
41. Palmer, M.; White, P. S. (1994). "On the existence of ecological communities" (PDF). Journal of Vegetation Sciences. 5 (2): 279–282. doi:10.2307/3236162. JSTOR 3236162.
42. Prentice; I. C.; Harrison, S. P.; Leemans, R.; Monserud, R. A.; Solomon, A. M. (1992). "Special paper: A global biome model based on plant physiology and dominance, soil properties and climate". Journal of Biogeography. 19 (2): 117–134. doi:10.2307/2845499. JSTOR 2845499.
43. Turnbaugh, Peter J.; Ley, Ruth E.; Hamady, Micah; Fraser-Liggett, Claire M.; Knight, Rob; Gordon, Jeffrey I. (2007). "The human microbiome project". Nature. 449 (7164): 804–810. Bibcode:2007Natur.449..804T. doi:10.1038/nature06244. PMC 3709439 . PMID 17943116.
44. DeLong, E. F. (2009). "The microbial ocean from genomes to biomes" (PDF). Nature. 459 (7244): 200–206. Bibcode:2009Natur.459..200D. doi:10.1038/nature08059. PMID 19444206.
45. Igamberdiev, Abir U.; Lea, P. J. (2006). "Land plants equilibrate O2 and CO2 concentrations in the atmosphere"(PDF). Photosynthesis Research. 87 (2): 177–194. doi:10.1007/s11120-005-8388-2. PMID 16432665.
46. Lovelock, J.; Margulis, Lynn (1973). "Atmospheric homeostasis by and for the biosphere: The Gaia hypothesis". Tellus. 26: 2–10. Bibcode:1974Tell...26....2L. doi:10.1111/j.2153-3490.1974.tb01946.x.
47. Lovelock, J. (2003). "The living Earth". Nature. 426 (6968): 769–770. Bibcode:2003Natur.426..769L. doi:10.1038/426769a. PMID 14685210.
48. Waples, R. S.; Gaggiotti, O. (2006). "What is a population? An empirical

evaluation of some genetic methods for identifying the number of gene pools and their degree of connectivity". Molecular Ecology. 15 (6): 1419–1439. doi:10.1111/j.1365-294X.2006.02890.x. PMID 16629801.

49. Turchin, P. (2001). "Does population ecology have general laws?". Oikos. 94 (1): 17–26. doi:10.1034/j.1600-0706.2001.11310.x.
50. Vandermeer, J. H.; Goldberg, D. E. (2003). Population Ecology: First Principles. Woodstock, Oxfordshire: Princeton University Press. ISBN 0-691-11440-4.
51. Berryman, A. A. (1992). "The origins and evolution of predator-prey theory". Ecology. 73 (5): 1530–1535. doi:10.2307/1940005. JSTOR 1940005.
52. Anderson, D. R.; Burnham, K. P.; Thompson, W. L. (2000). "Null hypotheses testing: Problems, prevalence, and an alternative" (PDF). J. Wildl. Mngt. 64 (4): 912–923.
53. Johnson, J. B.; Omland, K. S. (2004). "Model selection in ecology and evolution" (PDF). Trends in Ecology and Evolution. 19 (2): 101–108. doi:10.1016/j.tree.2003.10.013. PMID 16701236.
54. Levins, R. (1969). "Some demographic and genetic consequences of environmental heterogeneity for biological control". Bulletin of the Entomological Society of America. 15: 237–240. ISBN 978-0-231-12680-9.
55. Levins, R. (1970). "Extinction". In Gerstenhaber, M. Some Mathematical Questions in Biology. pp. 77–107. ISBN 978-0-8218-1152-8.
56. Smith, M. A.; Green, D. M. (2005). "Dispersal and the metapopulation paradigm in amphibian ecology and conservation: Are all amphibian populations metapopulations?". Ecography. 28 (1): 110–128. doi:10.1111/j.0906-7590.2005.04042.x.
57. Hanski, I. (1998). "Metapopulation dynamics" (PDF). Nature. 396 (6706): 41–49. Bibcode:1998Natur.396...41H. doi:10.1038/23876.
58. Nebel, S. (2010). "Animal migration". Nature Education Knowledge. 10 (1): 29.
59. Clark, J. S.; Fastie, C.; Hurtt, G.; Jackson, S. T.; Johnson, C.; King, G. A.; Lewis, M.; Lynch, J.; Pacala, S.; et al. (1998). "Reid's paradox of rapid plant migration" (PDF). BioScience. 48 (1): 13–24. doi:10.2307/1313224. JSTOR 1313224.
60. Dingle, H. (1996-01-18). Migration: The Biology of Life on the Move. Oxford University Press. p. 480. ISBN 0-19-509723-8.
61. Hanski, I.; Gaggiotti, O. E., eds. (2004). Ecology, Genetics and Evolution of Metapopulations. Burlington, MA: Elsevier Academic Press. ISBN 0-12-323448-4.
62. MacKenzie; D.I. (2006). Occupancy Estimation and Modeling: Inferring Patterns and Dynamics of Species Occurrence. London, UK: Elsevier Academic Press. p. 324. ISBN 978-0-12-088766-8.
63. Johnson, M. T.; Strinchcombe, J. R. (2007). "An emerging synthesis between community ecology and evolutionary biology". Trends in Ecology and Evolution. 22 (5): 250–257. doi:10.1016/j.tree.2007.01.014. PMID 17296244.
64. Tansley, A. G. (1935). "The use and abuse of vegetational concepts and terms"

(PDF). Ecology. 16 (3): 284–307. doi:10.2307/1930070. JSTOR 1930070.

65. Brinson, M. M.; Lugo, A. E.; Brown, S (1981). "Primary Productivity, Decomposition and Consumer Activity in Freshwater Wetlands". Annual Review of Ecology and Systematics. 12: 123–161. doi:10.1146/annurev.es.12.110181.001011.
66. Marsh, G. P. (1864). Man and Nature: Physical Geography as Modified by Human Action. Cambridge, MA: Belknap Press. p. 560.
67. O'Neil, R. V. (2001). "Is it time to bury the ecosystem concept? (With full military honors, of course!)" (PDF). Ecology. 82 (12): 3275–3284. doi:10.1890/0012-9658(2001)082[3275:IITTBT]2.0.CO;2. ISSN 0012-9658.
68. Levin, S. A. (1998). "Ecosystems and the biosphere as complex adaptive systems". Ecosystems. 1 (5): 431–436. CiteSeerX 10.1.1.83.6318 . doi:10.1007/s100219900037.
69. Pimm, S. (2002). Food Webs. University of Chicago Press. p. 258. ISBN 978-0-226-66832-1.
70. Pimm, S. L.; Lawton, J. H.; Cohen, J. E. (1991). "Food web patterns and their consequences" (PDF). Nature. 350 (6320): 669–674. Bibcode:1991Natur.350..669P. doi:10.1038/350669a0.
71. Worm, B.; Duffy, J. E. (2003). "Biodiversity, productivity and stability in real food webs". Trends in Ecology and Evolution. 18 (12): 628–632. doi:10.1016/j.tree.2003.09.003.
72. McCann, K. (2007). "Protecting biostructure" (PDF). Nature. 446 (7131): 29. Bibcode:2007Natur.446...29M. doi:10.1038/446029a. PMID 17330028.
73. Wilbur, H. W. (1997). "Experimental ecology of food webs: Complex systems in temporary ponds" (PDF). Ecology. 78 (8): 2279–2302. doi:10.1890/0012-9658(1997)078[2279:EEOFWC]2.0.CO;2. ISSN 0012-9658.
74. Emmerson, M.; Yearsley, J. M. (2004). "Weak interactions, omnivory and emergent food-web properties" (PDF). Philosophical Transactions of the Royal Society B. 271 (1537): 397–405. doi:10.1098/rspb.2003.2592.
75. Krause, A. E.; Frank, K. A.; Mason, D. M.; Ulanowicz, R. E.; Taylor, W. W. (2003). "Compartments revealed in food-web structure" (PDF). Nature. 426 (6964): 282–285. Bibcode:2003Natur.426..282K. doi:10.1038/nature02115. PMID 14628050.
76. Egerton, Frank N. (2007). "Understanding food chains and food webs, 1700–1970". Bulletin of the Ecological Society of America. 88: 50–69. doi:10.1890/0012-9623(2007)88[50:UFCAFW]2.0.CO;2. ISSN 0012-9623.
77. Shurin, J. B.; Gruner, D. S.; Hillebrand, H. (2006). "All wet or dried up? Real differences between aquatic and terrestrial food webs". Proceedings of the Royal Society B. 273 (1582): 1–9. doi:10.1098/rspb.2005.3377. PMC 1560001 . PMID 16519227.
78. Edwards, J.; Fraser, K. (1983). "Concept maps as reflectors of conceptual understanding". Research in Science Education. 13: 19–26. Bibcode:1983RScEd..13...19E. doi:10.1007/BF02356689.

79. Hairston Jr., N. G.; Hairston Sr., N. G. (1993). "Cause-effect relationships in energy flow, trophic structure, and interspecific interactions" (PDF). The American Naturalist. 142 (3): 379–411. doi:10.1086/285546.
80. Duffy, J. Emmett; Cardinale, Bradley J.; France, Kristin E.; McIntyre, Peter B.; Thébault, Elisa; Loreau, Michel (2007). "The functional role of biodiversity in ecosystems: incorporating trophic complexity". Ecology Letters. 10 (6): 522–538. doi:10.1111/j.1461-0248.2007.01037.x. PMID 17498151.
81. Elton, C. S. (1927). Animal Ecology. London, UK.: Sidgwick and Jackson. ISBN 0-226-20639-4.
82. Davic, R. D. (2003). "Linking keystone species and functional groups: a new operational definition of the keystone species concept". Conservation Ecology. 7 (1): r11.
83. Oksanen, L. (1991). "Trophic levels and trophic dynamics: A consensus emerging?". Trends in Ecology and Evolution. 6 (2): 58–60. doi:10.1016/0169-5347(91)90124-G. PMID 21232425.
84. Loehle, C.; Pechmann, Joseph H. K. (1988). "Evolution: The missing ingredient in systems ecology". The American Naturalist. 132 (9): 884–899. doi:10.1086/284895. JSTOR 2462267.
85. Ulanowicz, R. E.; Kemp, W. Michael (1979). "Toward canonical trophic aggregations". The American Naturalist. 114 (6): 871–883. doi:10.1086/283534. JSTOR 2460557.
86. Li, B. (2000). "Why is the holistic approach becoming so important in landscape ecology?". Landscape and Urban Planning. 50 (1–3): 27–41. doi:10.1016/S0169-2046(00)00078-5.
87. Polis, G. A.; Strong, D. R. (1996). "Food web complexity and community dynamics" (PDF). The American Naturalist. 147 (5): 813–846. doi:10.1086/285880.
88. Thompson, R. M.; Hemberg, M.; Starzomski, B. M.; Shurin, J. B. (2007). "Trophic levels and trophic tangles: The prevalence of omnivory in real food webs" (PDF). Ecology. 88 (3): 612–617. doi:10.1890/05-1454. PMID 17503589.
89. Fischer, J.; Lindenmayer, D. B.; Manning, A. D. (2006). "Biodiversity, ecosystem function, and resilience: ten guiding principles for commodity production landscapes" (PDF). Frontiers in Ecology and the Environment. 4 (2): 80–86. doi:10.1890/1540-9295(2006)004[0080:BEFART]2.0.CO;2. ISSN 1540-9295.
90. Libralato, S.; Christensen, V.; Pauly, D. (2006). "A method for identifying keystone species in food web models" (PDF). Ecological Modelling. 195 (3–4): 153–171. doi:10.1016/j.ecolmodel.2005.11.029.
91. Mills, L. S.; Soule, M. E.; Doak, D. F. (1993). "The keystone-species concept in ecology and conservation". BioScience. 43 (4): 219–224. doi:10.2307/1312122. JSTOR 1312122.
92. Anderson, P. K. (1995). "Competition, predation, and the evolution and extinction of Steller's sea cow, Hydrodamalis gigas". Marine Mammal Science. 11 (3): 391–394. doi:10.1111/j.1748-7692.1995.tb00294.x.

93. Polis, G. A.; Sears, Anna L. W.; Huxel, Gary R.; Strong, Donald R.; Maron, John (2000). "When is a trophic cascade a trophic cascade?" (PDF). Trends in Ecology and Evolution. 15 (11): 473–475. doi:10.1016/S0169-5347(00)01971-6. PMID 11050351.

94. Novikoff, A. B. (1945). "The concept of integrative levels and biology" (PDF). Science. 101 (2618): 209–215. Bibcode:1945Sci...101..209N. doi:10.1126/science.101.2618.209. PMID 17814095.

95. Schneider, D. D. (2001). "The rise of the concept of scale in ecology" (PDF). BioScience. 51 (7): 545–553. doi:10.1641/0006-3568(2001)051[0545:TROTCO]2.0.CO;2. ISSN 0006-3568.

96. Molnar, J.; Marvier, M.; Kareiva, P. (2004). "The sum is greater than the parts" (PDF). Conservation Biology. 18 (6): 1670–1671. doi:10.1111/j.1523-1739.2004.00l07.x.

97. Loehle, C. (2004). "Challenges of ecological complexity". Ecological Complexity. 1 (1): 3–6. doi:10.1016/j.ecocom.2003.09.001.

98. Odum, E. P. (1977). "The emergence of ecology as a new integrative discipline". Science. 195 (4284): 1289–1293. Bibcode:1977Sci...195.1289O. doi:10.1126/science.195.4284.1289. PMID 17738398.

99. Scheffer, M.; Carpenter, S.; Foley, J. A.; Walker, B.; Walker, B. (2001). "Catastrophic shifts in ecosystems" (PDF). Nature. 413 (6856): 591–596. Bibcode:2001Natur.413..591S. doi:10.1038/35098000. PMID 11595939.

100. "Welcome to ILTER — ILTER". International Long Term Ecological Research. Retrieved 2010-03-16.

101. Silverton, Jonathan; Poulton, Paul; Johnston, Edward; Edwards, Grant; Heard, Matthew; Biss, Pamela M. (2006). "The Park Grass Experiment 1856–2006: Its contribution to ecology" (PDF). Journal of Ecology. 94 (4): 801–814. doi:10.1111/j.1365-2745.2006.01145.x.

102. "Hubbard Brook Ecosystem Study Front Page". Retrieved 2010-03-16.

103. Liu, J.; Dietz, Thomas; Carpenter, Stephen R.; Folke, Carl; Alberti, Marina; Redman, Charles L.; Schneider, Stephen H.; Ostrom, Elinor; Pell, Alice N.; et al. (2009). "Coupled human and natural systems" (PDF). AMBIO: A Journal of the Human Environment. 36 (8): 639–649. doi:10.1579/0044-7447(2007)36[639:CHANS]2.0.CO;2. ISSN 0044-7447.

104. Mikkelson, G. M. (2010). "Part-whole relationships and the unity of ecology". In Skipper, R. A.; Allen, C.; Ankeny, R.; Craver, C. F.; Darden, L.; Richardson, R.C. Philosophy Across the Life Sciences (PDF). Cambridge, MA: MIT Press.

105. b Wilson, D. S. (1988). "Holism and reductionism in evolutionary ecology". Oikos. 53 (2): 269–273. doi:10.2307/3566073. JSTOR 3566073.

106. Miles, D. B.; Dunham, A. E. (1993). "Historical perspectives in ecology and evolutionary biology: The use of phylogenetic comparative analyses". Annual Review of Ecology and Systematics. 24: 587–619. doi:10.1146/annurev.es.24.110193.003103.

107. Craze, P., ed. (2 August 2012). "Trends in Ecology and Evolution". Cell Press, Elsevier, Inc.

108. Allee, W. C.; Park, O.; Emerson, A. E.; Park, T.; Schmidt, K. P. (1949). Principles of Animal Ecology. W. B. Sunders, Co. p. 837. ISBN 0-7216-1120-6.

109. Rickleffs, Robert, E. (1996). The Economy of Nature. University of Chicago Press. p. 678. ISBN 0-7167-3847-3.

110. Yoshida, T. "Rapid evolution drives ecological dynamics in a predator–prey system". Nature Publishing Group.

111. Stuart-Fox, D.; Moussalli, A. (2008). "Selection for social signalling drives the evolution of chameleon colour change". PLoS Biology. 6 (1): e25. doi:10.1371/journal.pbio.0060025. PMC 2214820 . PMID 18232740.

112. Karban, R. (2008). "Plant behaviour and communication". Ecology Letters. 11 (7): 727–739. doi:10.1111/j.1461-0248.2008.01183.x. PMID 18400016.

113. Tinbergen, N. (1963). "On aims and methods of ethology" (PDF). Zeitschrift für Tierpsychologie. 20 (4): 410–433. doi:10.1111/j.1439-0310.1963.tb01161.x.

114. Hamner, W. M. (1985). "The importance of ethology for investigations of marine zooplankton". Bulletin of Marine Science. 37 (2): 414–424.

115. Strassmann, J. E.; Zhu, Y.; Queller, D. C. (2000). "Altruism and social cheating in the social amoeba Dictyostelium discoideum". Nature. 408 (6815): 965–967. doi:10.1038/35050087. PMID 11140681.

116. Sakurai, K. (1985). "An attelabid weevil (Euops splendida) cultivates fungi". Journal of Ethology. 3 (2): 151–156. doi:10.1007/BF02350306.

117. Anderson, J. D. (1961). "The courtship behaviour of Ambystoma macrodactylum croceum". Copeia. 2 (2): 132–139. doi:10.2307/1439987. JSTOR 1439987.

118. "Behavioral Ecology". International Society for Behavioral Ecology. Retrieved 15 April 2011.

119. Gould, Stephen J.; Vrba, Elizabeth S. (1982). "Exaptation–a missing term in the science of form". Paleobiology. 8 (1): 4–15.

120. Wilson, E. O. (2000). Sociobiology: The New Synthesis (25th anniversary ed.). President and Fellows of Harvard College. ISBN 978-0-674-00089-6. Cite error: Invalid <ref> tag; name "Wilson00" defined multiple times with different content (see thehelp page).

121. Ives, A. R.; Cardinale, B. J.; Snyder, W. E. (2004). "A synthesis of subdisciplines: Predator–prey interactions, and biodiversity and ecosystem functioning" (PDF). Ecology Letters. 8 (1): 102–116. doi:10.1111/j.1461-0248.2004.00698.x.

122. Krebs, J. R.; Davies, N. B. (1993). An Introduction to Behavioural Ecology. Wiley-Blackwell. p. 432. ISBN 978-0-632-03546-5.

123. Webb, J. K.; Pike, D. A.; Shine, R. (2010). "Olfactory recognition of predators by nocturnal lizards: safety outweighs thermal benefits". Behavioural Ecology. 21 (1): 72–77. doi:10.1093/beheco/arp152.

124. Cooper, W. E.; Frederick, W. G. (2010). "Predator lethality, optimal escape

behavior, and autotomy". Behavioral Ecology. 21 (1): 91–96. doi:10.1093/beheco/arp151.

125. Eastwood, R. (2004). "Successive replacement of tending ant species at aggregations of scale insects (Hemiptera: Margarodidae and Eriococcidae) on Eucalyptus in south-east Queensland" (PDF). Australian Journal of Entomology. 43: 1–4. doi:10.1111/j.1440-6055.2003.00371.x.

126. Kodric-Brown, A.; Brown, J. H. (1984). "Truth in advertising: The kinds of traits favored by sexual selection" (PDF). The American Naturalist. 124 (3): 309–323. doi:10.1086/284275.[dead link]

127. Adrian G Palacios, Francisco Bozinovic; Bozinovic (2003). "An "enactive" approach to integrative and comparative biology: Thoughts on the table" (PDF). Biology Research. 36 (1): 95–99. doi:10.4067/S0716-97602003000100008. PMID 12795209.

128. Reuven Dukas (1998). "§1.3 Why study cognitive ecology?". In Reuven Dukas. Cognitive Ecology: The Evolutionary Ecology of Information Processing and Decision Making. University of Chicago Press. p. 4. ISBN 9780226169323.

129. Reuven Dukas; John M. Ratcliffe (2009). "Introduction". In Reuven Dukas; John M. Ratcliffe. Cognitive Ecology II. University of Chicago Press. pp. 1 ff. ISBN 9780226169378. Cognitive ecology focuses on he ecology and evolution of "cognition" defined as the neuronal processes concerned with the acquisition, retention, and use of information....we ought to rely on ecological and evolutionary knowledge for studying cognition.

130. Francisco J Varela; Evan Thompson; Eleanor Rosch (1993). The Embodied Mind: Cognitive Science and Human Experience (Paperback ed.). MIT Press. p. 174. ISBN 9780262261234.

131. Sherman, P. W.; Lacey, E. A.; Reeve, H. K.; Keller, L. (1995). "The eusociality continuum" (PDF). Behavioural Ecology. 6 (1): 102–108. doi:10.1093/beheco/6.1.102.

132. Wilson, D. S.; Wilson, E. O. (2007). "Rethinking the theoretical foundation of sociobiology". The Quarterly Review of Biology. 82 (4): 327–348. doi:10.1086/522809. PMID 18217526.

133. Page, R. D. M. (1991). "Clocks, clades, and cospeciation: Comparing rates of evolution and timing of cospeciation events in host-parasite assemblages". Systematic Zoology. 40 (2): 188–198. doi:10.2307/2992256. JSTOR 2992256.

134. Herre, E. A.; Knowlton, N.; Mueller, U. G.; Rehner, S. A. (1999). "The evolution of mutualisms: exploring the paths between conflict and cooperation" (PDF). Trends in Ecology and Evolution. 14 (2): 49–53. doi:10.1016/S0169-5347(98)01529-8. PMID 10234251.

135. Gilbert, F. S. (1990). Insect life cycles: Genetics, evolution, and co-ordination. New York, NY: Springer-Verlag. p. 258. ISBN 0-387-19550-5.

136. Kiers, E. T.; van der Heijden, M. G. A. (2006). "Mutualistic stability in the arbuscular mycorrhizal symbiosis: Exploring hypotheses of evolutionary cooperation" (PDF). Ecology. 87 (7): 1627–1636. doi:10.1890/0012-

9658(2006)87[1627:MSITAM]2.0.CO;2. ISSN 0012-9658. PMID 16922314.

137. Strain, B. R. (1985). "Physiological and ecological controls on carbon sequestering in terrestrial ecosystems". Biogeochemistry. 1 (3): 219–232. doi:10.1007/BF02187200.

138. Bronstein, J. L. (2001). "The exploitation of mutualisms". Ecology Letters. 4 (3): 277–287. doi:10.1046/j.1461-0248.2001.00218.x.

139. Irwin, Rebecca E.; Bronstein, Judith L.; Manson, Jessamyn S.; Richardson, Leif (2010). "Nectar robbing: Ecological and evolutionary perspectives". Annual Review of Ecology, Evolution, and Systematics. 41 (2): 271–292. doi:10.1146/annurev.ecolsys.110308.120330.

140. Boucher, D. H.; James, S.; Keeler, K. H. (1982). "The ecology of mutualism". Annual Review of Ecology and Systematics. 13: 315–347. doi:10.1146/annurev.es.13.110182.001531.

141. King, K. C.; Delph, L. F.; Jokela, J.; Lively, C. M. (2009). "The geographic mosaic of sex and the Red Queen". Current Biology. 19 (17): 1438–1441. doi:10.1016/j.cub.2009.06.062. PMID 19631541.

142. Parenti, L. R.; Ebach, M. C. (2009). Comparative Biogeography: Discovering and Classifying Biogeographical Patterns of a Dynamic Earth. London, England: University of California Press. ISBN 978-0-520-25945-4.

143. "Journal of Biogeography". Wiley. Retrieved 3 August 2012.

144. MacArthur, R.; Wilson, E. O. (1967). "The Theory of Island Biogeography". Princeton, NJ: Princeton University Press.

145. Wiens, J. J.; Donoghue, M. J. (2004). "Historical biogeography, ecology and species richness" (PDF). Trends in Ecology and Evolution. 19 (12): 639–644. doi:10.1016/j.tree.2004.09.011. PMID 16701326.

146. Morrone, J. J.; Crisci, J. V. (1995). "Historical biogeography: Introduction to methods". Annual Review of Ecology and Systematics. 26: 373–401. doi:10.1146/annurev.es.26.110195.002105.

147. Svenning, Jens-Christian; Condi, R. (2008). "Biodiversity in a warmer world". Science. 322 (5899): 206–207. doi:10.1126/science.1164542. PMID 18845738.

148. Landhäusser, Simon M.; Deshaies, D.; Lieffers, V. J. (2009). "Disturbance facilitates rapid range expansion of aspen into higher elevations of the Rocky Mountains under a warming climate". Journal of Biogeography. 37 (1): 68–76. doi:10.1111/j.1365-2699.2009.02182.x.

149. Reznick, D.; Bryant, M. J.; Bashey, F. (2002). "r- and K-selection revisited: The role of population regulation in life-history evolution" (PDF). Ecology. 83 (6): 1509–1520. doi:10.1890/0012-9658(2002)083[1509:RAKSRT]2.0.CO;2. ISSN 0012-9658.

150. Pianka, E. R. (1972). "r and K Selection or b and d Selection?". The American Naturalist. 106 (951): 581–588. doi:10.1086/282798.

151. Rieseberg, L. (ed.). "Molecular Ecology". Molecular Ecology. Wiley. doi:10.1111/(ISSN)1365-294X.

152. Avise, J. (1994). Molecular Markers, Natural History and Evolution. Kluwer Academic Publishers. ISBN 0-412-03771-8.
153. O'Brian, E.; Dawson, R. (2007). "Context-dependent genetic benefits of extra-pair mate choice in a socially monogamous passerine" (PDF). Behavioral Ecology and Sociobiology. 61 (5): 775–782. doi:10.1007/s00265-006-0308-8.
154. Avise, J. (2000). Phylogeography: The History and Formation of Species. President and Fellows of Harvard College. ISBN 0-674-66638-0.
155. Rachel Carson (1962). ""Silent Spring" (excerpt)". Houghton Miffin. Retrieved 4 October 2012.
156. Young, G. L. (1974). "Human ecology as an interdisciplinary concept: A critical inquiry". Advances in Ecological Research. Advances in Ecological Research. 8: 1–105. doi:10.1016/S0065-2504(08)60277-9. ISBN 978-0-12-013908-8.
157. Gross, M. (2004). "Human geography and ecological sociology: the unfolding of human ecology, 1890 to 1930 – and beyond". Social Science History. 28 (4): 575–605. doi:10.1215/01455532-28-4-575.
158. "Millennium Ecosystem Assessment – Synthesis Report". United Nations. 2005. Retrieved 4 February 2010.
159. de Groot, R. S.; Wilson, M. A.; Boumans, R. M. J. (2002). "A typology for the classification, description and valuation of ecosystem functions, goods and services" (PDF). Ecological Economics. 41 (3): 393–408. doi:10.1016/S0921-8009(02)00089-7.
160. Aguirre, A. A. (2009). "Biodiversity and human health". EcoHealth. 6: 153–156. doi:10.1007/s10393-009-0242-0.
161. Grumbine, R. E. (1994). "What is ecosystem management?" (PDF). Conservation Biology. 8 (1): 27–38. doi:10.1046/j.1523-1739.1994.08010027.x.
162. Wilson, E. O. (1992). The Diversity of Life. Harvard University Press. p. 440. ISBN 978-0-674-05817-0.
163. Slocombe, D. S. (1993). "Implementing ecosystem-based management". BioScience. 43 (9): 612–622. doi:10.2307/1312148. JSTOR 1312148.
164. Hobss, R. J.; Harris, J. A. (2001). "Restoration ecology: Repairing the Earth's ecosystems in the new millennium"(PDF). Restoration Ecology. 9 (2): 239–246. doi:10.1046/j.1526-100x.2001.009002239.x.
165. Mason, H. L.; Langenheim, J. H. (1957). "Language analysis and the concept "environment"". Ecology. 38 (2): 325–340. doi:10.2307/1931693. JSTOR 1931693.
166. Kleese, D. A. (2001). "Nature and nature in Psychology". Journal of Theoretical and Philosophical Psychology. 21: 61–79. doi:10.1037/h0091199.
167. Campbell, Neil A.; Williamson, Brad; Heyden, Robin J. (2006). Biology: Exploring Life. Boston, Massachusetts: Pearson Prentice Hall. ISBN 0-13-250882-6.
168. Kormondy, E. E. (1995). Concepts of Ecology (4th ed.). Benjamin Cummings. ISBN 0-13-478116-3.
169. Hughes, A. R. "Disturbance and diversity: an ecological chicken and egg

problem". Nature Education Knowledge. 1 (8): 26.

170. Levin, S. A. (1992). "The problem of pattern and scale in ecology: The Robert H. MacArthur Award" (PDF). Ecology. 73 (6): 1943–1967. doi:10.2307/1941447. JSTOR 1941447. Retrieved 2010-03-16.
171. ^ Jump up to:a b Holling, C. S. (1973). "Resilience and stability of ecological systems". Annual Review of Ecology and Systematics. 4 (1): 1–23. doi:10.1146/annurev.es.04.110173.000245. JSTOR 2096802.
172. ^ Jump up to:a b Folke, C.; Carpenter, S.; Walker, B.; Scheffer, M.; Elmqvist, T.; Gunderson, L.; Holling, C.S. (2004). "Regime shifts, resilience, and biodiversity in ecosystem management" (PDF). Annual Review of Ecology and Systematics. 35: 557–581. doi:10.1146/annurev.ecolsys.35.021103.105711. JSTOR 2096802.
173. Morgan Ernest, S. K.; Enquist, Brian J.; Brown, James H.; Charnov, Eric L.; Gillooly, James F.; Savage, Van M.; White, Ethan P.; Smith, Felisa A.; Hadly, Elizabeth A.; Haskell, John P.; Lyons, S. Kathleen; Maurer, Brian A.; Niklas, Karl J.; Tiffney, Bruce (2003). "Thermodynamic and metabolic effects on the scaling of production and population energy use"(PDF). Ecology Letters. 6 (11): 990–995. doi:10.1046/j.1461-0248.2003.00526.x.
174. Allègre, Claude J.; Manhès, Gérard; Göpel, Christa (1995). "The age of the Earth". Geochimica et Cosmochimica Acta. 59 (8): 1455–1456. Bibcode:1995GeCoA..59.1445A. doi:10.1016/0016-7037(95)00054-4.
175. Wills, C.; Bada, J. (2001). The Spark of Life: Darwin and the Primeval Soup. Cambridge, MA: Perseus Publishing. ISBN 978-0-7382-0493-2.
176. Goldblatt, Colin; Lenton, Timothy M.; Watson, Andrew J. (2006). "Bistability of atmospheric oxygen and the Great Oxidation" (PDF). Nature. 443 (7112): 683–686. Bibcode:2006Natur.443..683G. doi:10.1038/nature05169. PMID 17036001.
177. Catling, D. C.; Claire, M. W. (2005). "How Earth's atmosphere evolved to an oxic state: A status report"(PDF). Earth and Planetary Science Letters. 237: 1–20. Bibcode:2005E&PSL.237....1C. doi:10.1016/j.epsl.2005.06.013.
178. Cronk, J. K.; Fennessy, M. S. (2001). Wetland Plants: Biology and Ecology. Washington, D.C.: Lewis Publishers. ISBN 1-56670-372-7.
179. Evans, D. H.; Piermarini, P. M.; Potts, W. T. W. (1999). "Ionic transport in the fish gill epithelium" (PDF). Journal of Experimental Zoology. 283 (7): 641–652. doi:10.1002/(SICI)1097-010X(19990601)283:7<641::AID-JEZ3>3.0.CO;2-W.
180. Swenson, N. G.; Enquist, B. J. (2008). "The relationship between stem and branch wood specific gravity and the ability of each measure to predict leaf area". American Journal of Botany. 95 (4): 516–519. doi:10.3732/ajb.95.4.516. PMID 21632377.
181. Gartner, Gabriel E.A.; Hicks, James W.; Manzani, Paulo R.; et al. (2010). "Phylogeny, ecology, and heart position in snakes" (PDF). Physiological and Biochemical Zoology. 83 (1): 43–54. doi:10.1086/648509. PMID 19968564.
182. Jacobsen, D. (2008). "Low oxygen pressure as a driving factor for the altitudinal decline in taxon richness of stream macroinvertebrates". Oecologia. 154 (4):

795–807. doi:10.1007/s00442-007-0877-x. PMID 17960424.

183. Wheeler, T. D.; Stroock, A. D. (2008). "The transpiration of water at negative pressures in a synthetic tree". Nature. 455 (7210): 208–212. Bibcode:2008Natur.455..208W. doi:10.1038/nature07226. PMID 18784721.
184. Pockman, W. T.; Sperry, J. S.; O'Leary, J. W. (1995). "Sustained and significant negative water pressure in xylem". Nature. 378 (6558): 715–716. Bibcode:1995Natur.378..715P. doi:10.1038/378715a0.
185. Zimmermann, U.; Schneider, H.; Wegner, L. H.; Wagner, M.; Szimtenings, A.; Haase, F.; Bentrup, F. W. (2002). "What are the driving forces for water lifting in the xylem conduit?". Physiologia Plantarum. 114 (3): 327–335. doi:10.1034/j.1399-3054.2002.1140301.x. PMID 12060254.
186. Kastak, D.; Schusterman, R. J. (1998). "Low-frequency amphibious hearing in pinnipeds: Methods, measurements, noise, and ecology". Journal of the Acoustical Society of America. 103 (4): 2216–2228. Bibcode:1998ASAJ..103.2216K. doi:10.1121/1.421367. PMID 9566340.
187. Nishiguchi, Y.; Ito, I.; Okada, M. (2010). "Structure and function of lactate dehydrogenase from hagfish". Marine Drugs. 8 (3): 594–607. doi:10.3390/md8030594. PMC 2857353 . PMID 20411117.
188. Friedman, J.; Harder, L. D. (2004). "Inflorescence architecture and wind pollination in six grass species"(PDF). Functional Ecology. 18 (6): 851–860. doi:10.1111/j.0269-8463.2004.00921.x.
189. Harder, L. D.; Johnson, S. D. (2009). "Darwin's beautiful contrivances: evolutionary and functional evidence for floral adaptation" (PDF). New Phytologist. 183 (3): 530–545. doi:10.1111/j.1469-8137.2009.02914.x. PMID 19552694.
190. Shimeta, J.; Jumars, P. A.; Lessard, E. J. (1995). "Influences of turbulence on suspension feeding by planktonic protozoa; experiments in laminar shear fields" (PDF). Limnolology and Oceanography. 40 (5): 845–859. doi:10.4319/lo.1995.40.5.0845.
191. Etemad-Shahidi, A.; Imberger, J. (2001). "Anatomy of turbulence in thermally stratified lakes" (PDF). Limnolology and Oceanography. 46 (5): 1158–1170. doi:10.4319/lo.2001.46.5.1158.
192. Wolf, B. O.; Walsberg, G. E. (2006). "Thermal effects of radiation and wind on a small bird and implications for microsite selection". Ecology. 77 (7): 2228–2236. doi:10.2307/2265716. JSTOR 2265716.
193. Daubenmire, R. (1975). "Floristic plant geography of eastern Washington and northern Idaho". Journal of Biogeography. 2 (1): 1–18. doi:10.2307/3038197. JSTOR 3038197.
194. Steele, C. A.; Carstens, B. C.; Storfer, A.; Sullivan, J. (2005). "Testing hypotheses of speciation timing in Dicamptodon copei and Dicamptodon aterrimus (Caudata: Dicamptodontidae)" (PDF). Molecular Phylogenetics and Evolution. 36 (1): 90–100. doi:10.1016/j.ympev.2004.12.001. PMID 15904859.

195. Lenton, T. M.; Watson, A. (2000). "Redfield revisited. 2. What regulates the oxygen content of the atmosphere"(PDF). Global Biogeochemical Cycles. 14 (1): 249–268. Bibcode:2000GBioC..14..249L. doi:10.1029/1999GB900076.
196. Lobert, J. M.; Warnatz, J. (1993). "Emissions from the combustion process in vegetation". In Crutzen, P. J.; Goldammer, J. G. Fire in the Environment: The Ecological, Atmospheric and Climatic Importance of Vegetation Fires(PDF). Wiley. ISBN 978-0-471-93604-6.
197. Garren, K. H. (1943). "Effects of fire on vegetation of the southeastern United States". Botanical Review. 9 (9): 617–654. doi:10.1007/BF02872506.
198. Cooper, C. F. (1960). "Changes in vegetation, structure, and growth of southwestern pine forests since white settlement". Ecological Monographs. 30 (2): 130–164. doi:10.2307/1948549.
199. Cooper, C. F. (1961). "The ecology of fire". Scientific American. 204 (4): 150–160. doi:10.1038/scientificamerican0461-150.
200. van Wagtendonk, Jan W. (2007). "History and evolution of wildland fire use" (PDF). Fire Ecology Special Issue. 3 (2): 3–17. doi:10.4996/fireecology.0302003.
201. Boerner, R. E. J. (1982). "Fire and nutrient cycling in temperate ecosystems". BioScience. 32 (3): 187–192. doi:10.2307/1308941. JSTOR 1308941.
202. Goubitz, S.; Werger, M. J. A.; Ne'eman, G. (2009). "Germination response to fire-related factors of seeds from non-serotinous and serotinous cones". Plant Ecology. 169 (2): 195–204. doi:10.1023/A:1026036332277.
203. Ne'eman, G.; Goubitz, S.; Nathan, R. (2004). "Reproductive traits of Pinus halepensis in the light of fire: a critical review". Plant Ecology. 171 (1/2): 69–79. doi:10.1023/B:VEGE.0000029380.04821.99.
204. Flematti, Gavin R.; Ghisalberti, Emilio L.; Dixon, Kingsley W.; Trengove, R. D. (2004). "A compound from smoke that promotes seed germination" (PDF). Science. 305 (5686): 977. doi:10.1126/science.1099944. PMID 15247439.
205. Coleman, D. C.; Corssley, D. A.; Hendrix, P. F. (2004). Fundamentals of Soil Ecology (2nd ed.). Academic Press. ISBN 0-12-179726-0.
206. ^ Jump up to:a b Wilkinson, M. T.; Richards, P. J.; Humphreys, G. S. (2009). "Breaking ground: Pedological, geological, and ecological implications of soil bioturbation" (PDF). Earth-Science Reviews. 97 (1–4): 257–272. Bibcode:2009ESRv...97..257W. doi:10.1016/j.earscirev.2009.09.005.
207. Phillips, J. D. (2009). "Soils as extended composite phenotypes". Geoderma. 149 (1–2): 143–151. doi:10.1016/j.geoderma.2008.11.028.
208. Reinhardt, L.; Jerolmack, D.; Cardinale, B. J.; Vanacker, V.; Wright, J. "Dynamic interactions of life and its landscape: Feedbacks at the interface of geomorphology and ecology" (PDF). Earth Surf. Process. Landforms. 35: 78–101. Bibcode:2010ESPL...35...78R. doi:10.1002/esp.1912.
209. Davic, R. D.; Welsh, H. H. (2004). "On the ecological role of salamanders" (PDF). Annual Review of Ecology and Systematics. 35: 405–434. doi:10.1146/annurev.ecolsys.35.112202.130116.

210. Hasiotis, S. T. (2003). "Complex ichnofossils of solitary and social soil organisms: Understanding their evolution and roles in terrestrial paleoecosystems". Palaeogeography, Palaeoclimatology, Palaeoecology. 192 (2): 259–320. doi:10.1016/S0031-0182(02)00689-2.

211. Falkowski, P. G.; Fenchel, T.; Delong, E. F. (2008). "The microbial engines that drive Earth's biogeochemical cycles". Science. 320 (5879): 1034–1039. Bibcode:2008Sci...320.1034F. doi:10.1126/science.1153213. PMID 18497287.

212. Grace, J. (2004). "Understanding and managing the global carbon cycle". Journal of Ecology. 92 (2): 189–202. doi:10.1111/j.0022-0477.2004.00874.x.

213. Pearson, P. N.; Palmer, M. R. (2000). "Atmospheric carbon dioxide concentrations over the past 60 million years"(PDF). Nature. 406 (6797): 695–699. doi:10.1038/35021000. PMID 10963587.

214. Pagani, M.; Zachos, J. C.; Freeman, K. H.; Tipple, B.; Bohaty, S. (2005). "Marked decline in atmospheric carbon dioxide concentrations during the Paleogene". Science. 309 (5734): 600–603. Bibcode:2005Sci...309..600P. doi:10.1126/science.1110063. PMID 15961630.

215. Zhuan, Q.; Melillo, J. M.; McGuire, A. D.; Kicklighter, D. W.; Prinn, R. G.; Steudler, P. A.; Felzer, B. S.; Hu, S. (2007). "Net emission of CH4 and CO2 in Alaska: Implications for the region's greenhouse gas budget" (PDF). Ecological Applications. 17 (1): 203–212. doi:10.1890/1051-0761(2007)017[0203:NEOCAC]2.0.CO;2. ISSN 1051-0761. PMID 17479846.

216. Cox, Peter M.; Betts, Richard A.; Jones, Chris D.; Spall, Steven A.; Totterdell, Ian J. (2000). "Acceleration of global warming due to carbon-cycle feedbacks in a coupled climate model" (PDF). Nature. 408 (6809): 184–187. doi:10.1038/35041539. PMID 11089968.

217. Erwin, D. H. (2009). "Climate as a driver of evolutionary change". Current Biology. 19 (14): R575–R583. doi:10.1016/j.cub.2009.05.047. PMID 19640496.

218. Bamber, J. (2012). "Shrinking glaciers under scrutiny"(PDF). Nature. 482 (7386): 482–483. Bibcode:2012Natur.482..482B. doi:10.1038/nature10948. PMID 22318516.

219. Heimann, Martin; Reichstein, Markus (2008). "Terrestrial ecosystem carbon dynamics and climate feedbacks" (PDF). Nature. 451 (7176): 289–292. Bibcode:2008Natur.451..289H. doi:10.1038/nature06591. PMID 18202646.

220. Davidson, Eric A.; Janssens, Ivan A. (2006). "Temperature sensitivity of soil carbon decomposition and feedbacks to climate change". Nature. 440 (7081): 165–173. Bibcode:2006Natur.440..165D. doi:10.1038/nature04514. PMID 16525463.

221. ^ Jump up to:a b Egerton, F. N. (2001). "A history of the ecological sciences: early Greek origins" (PDF). Bulletin of the Ecological Society of America. 82 (1): 93–97.

222. ^ Jump up to:a b Benson, Keith R. (2000). "The emergence of ecology from natural history". Endeavour. 24 (2): 59–62. doi:10.1016/S0160-9327(99)01260-0. PMID 10969480.

223. Sober, E. (1980). "Evolution, population thinking, and essentialism". Philosophy of Science. 47 (3): 350–383. doi:10.1086/288942. JSTOR 186950.

224. Hughes, J. D. (1985). "Theophrastus as ecologist". Environmental Review. 9 (4): 296–306. doi:10.2307/3984460. JSTOR 3984460.

225. Hughes, J. D. (1975). "Ecology in ancient Greece". Inquiry. 18 (2): 115–125. doi:10.1080/00201747508601756.

226. Kingsland, S. (2004). "Conveying the intellectual challenge of ecology: An historical perspective" (PDF). Frontiers in Ecology and the Environment. 2 (7): 367–374. doi:10.1890/1540-9295(2004)002[0367:CTICOE]2.0.CO;2. ISSN 1540-9295.

227. Rosenzweig, M. L. (2003). "Reconciliation ecology and the future of species diversity" (PDF). Oryx. 37 (2): 194–205. doi:10.1017/s0030605303000371.

228. Hawkins, B. A. (2001). "Ecology's oldest pattern". Endeavor. 25 (3): 133. doi:10.1016/S0160-9327(00)01369-7.

229. McIntosh, R. P. (1985). The Background of Ecology: Concept and Theory. Cambridge University Press. p. 400. ISBN 0-521-27087-1.

230. Stauffer, R. C. (1957). "Haeckel, Darwin and ecology". The Quarterly Review of Biology. 32 (2): 138–144. doi:10.1086/401754.

231. Friederichs, K. (1958). "A definition of ecology and some thoughts about basic concepts". Ecology. 39 (1): 154–159. doi:10.2307/1929981. JSTOR 1929981.

232. Hinchman, L. P.; Hinchman, S. K. (2007). "What we owe the Romantics". Environmental Values. 16 (3): 333–354. doi:10.3197/096327107X228382.

233. Goodland, R. J. (1975). "The tropical origin of ecology: Eugen Warming's jubilee". Oikos. 26 (2): 240–245. doi:10.2307/3543715. JSTOR 3543715.

234. Egerton, F. N. (2007). "A history of the ecological sciences, part 23: Linnaeus and the economy of nature". Bulletin of the Ecological Society of America. 88 (1): 72–88. doi:10.1890/0012-9623(2007)88[72:AHOTES]2.0.CO;2. ISSN 0012-9623.

235. Kormandy, E. J.; Wooster, Donald (1978). "Review: Ecology/economy of nature—synonyms?". Ecology. 59 (6): 1292–1294. doi:10.2307/1938247. JSTOR 1938247.

236. Hector, A.; Hooper, R. (2002). "Darwin and the first ecological experiment". Science. 295 (5555): 639–640. doi:10.1126/science.1064815. PMID 11809960.

237. Sinclair, G. (1826). "On cultivating a collection of grasses in pleasure-grounds or flower-gardens, and on the utility of studying the Gramineae". London Gardener's Magazine. 1. New-Street-Square: A. & R. Spottiswoode. p. 115.

238. May, R. (1999). "Unanswered questions in ecology". Philosophical Transactions of the Royal Society B. 354 (1392): 1951–1959. doi:10.1098/rstb.1999.0534. PMC 1692702 . PMID 10670015.

239. Darwin, Charles (1859). On the Origin of Species (1st ed.). London, UK: John Murray. p. 1. ISBN 0-8014-1319-2.

240. Meysman, F. J. R.; Middelburg, Jack J.; Heip, C. H. R. (2006). "Bioturbation:

A fresh look at Darwin's last idea". Trends in Ecology and Evolution. 21 (22): 688–695. doi:10.1016/j.tree.2006.08.002. PMID 16901581.

241. Acot, P. (1997). "The Lamarckian cradle of scientific ecology". Acta Biotheoretica. 45 (3–4): 185–193. doi:10.1023/A:1000631103244.

242. Forbes, S. (1887). "The lake as a microcosm" (PDF). Bulletin of the Scientific Association. Peoria, IL: 77–87.

243. Hunt, Caroline Louisa (1912). The life of Ellen H. Richards (1st ed.). Boston: Whitcomb & Barrows.

244. Clements, F. E. (1905). Research methods in ecology. Lincoln, Neb.: University Pub. Comp. ISBN 0-405-10381-6.

245. Simberloff, D. (1980). "A succession of paradigms in ecology: Essentialism to materialism and probalism". Synthese. 43: 3–39. doi:10.1007/BF00413854.

246. Gleason, H. A. (1926). "The individualistic concept of the plant association" (PDF). Bulletin of the Torrey Botanical Club. 53 (1): 7–26. doi:10.2307/2479933. JSTOR 2479933.

247. Foster, J. B.; Clark, B. (2008). "The sociology of ecology: ecological organicism versus ecosystem ecology in the social construction of ecological science, 1926–1935" (PDF). Organization & Environment. 21 (3): 311–352. doi:10.1177/1086026608321632.

248. Allee, W. C. (1932). Animal Life and Social Growth. Baltimore: The Williams & Wilkins Company and Associates.

249. Cook, R. E. (1977). "Raymond Lindeman and the trophic-dynamic concept in ecology" (PDF). Science. 198 (4312): 22–26. Bibcode:1977Sci...198...22C. doi:10.1126/science.198.4312.22. PMID 17741875.

250. Odum, E. P. (1968). "Energy flow in ecosystems: A historical review". American Zoologist. 8 (1): 11–18. doi:10.1093/icb/8.1.11. JSTOR 3881528.

251. Ghilarov, A. M. (1995). "Vernadsky's biosphere concept: an historical perspective". The Quarterly Review of Biology. 70 (2): 193–203. doi:10.1086/418982. JSTOR 3036242.

252. Itô, Y. (1991). "Development of ecology in Japan, with special reference to the role of Kinji Imanishi". Journal of Ecological Research. 6 (2): 139–155. doi:10.1007/BF02347158.

253. Carson, R. (2002). Silent Spring. Houghton Mifflin Company. p. 348. ISBN 0-618-24906-0.

254. Palamar, C. R. (2008). "The justice of ecological restoration: Environmental history, health, ecology, and justice in the United States" (PDF). Human Ecology Review. 15 (1): 82–94.

255. Krebs, J. R.; Wilson, J. D.; Bradbury, R. B.; Siriwardena, G. M. (1999). "The second Silent Spring" (PDF). Nature. 400 (6745): 611–612. Bibcode:1999Natur.400..611K. doi:10.1038/23127.

CHAPTER 2

Study of Allometry

Chapter Contents

- **Introduction**
- **Isometric Scaling and Geometric Similarity**
- **Allometric Scaling**
- **Determining if a System is Scaling with Allometry**
- **Physiological Scaling**
- **Allometric Scaling in Fluid Locomotion**
- **Allometric Engineering**
- **Allometry in Characteristics of a City**
- **Examples**
- **Determinants of Size in Different Species**
- **Alternative Stable State**

Allometry, also called biological scaling, is the change in organisms in relation to proportional changes in body size. An example of allometry can be seen in mammals. Ranging from the mouse to the elephant, as the body gets larger, in general hearts beat more slowly, brains get bigger, bones get proportionally shorter and thinner, and life spans lengthen. Even ecologically flexible characteristics, such as population density and the size of home ranges, scale in a predictive way with body size.

INTRODUCTION

Allometry is the study of the relationship of body size to shape,anatomy, physiology and finally behavior. It was first outlined by Otto Snell in 1892, by D Arcy Thompson in 1917 in *On Growth and Form* and by Julian Huxley in 1932. It is a well-known field of study, particularly in statistical shape analysis for its theoretical developments, as well as in biology for practical applications to the differential growth rates of the parts of a living organism's body. One application is in the study of various insect species (e.g., the Hercules Beetle), where a small change in overall body size can lead to an enormous and disproportionate increase in the dimensions of appendages such as legs, antennae, or horns. The relationship between the two measured quantities is often expressed as a power law equation:

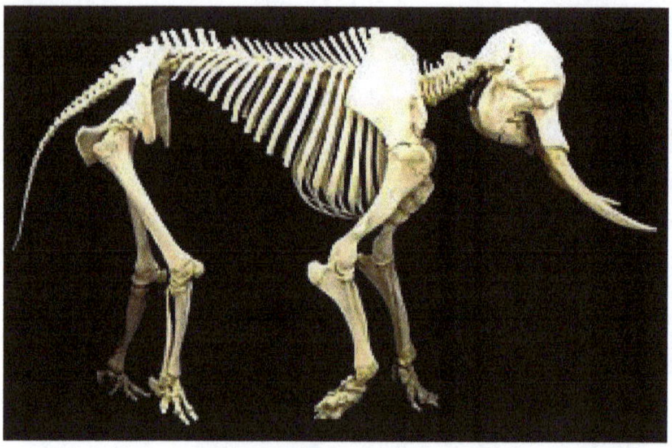

Skeleton of an elephant

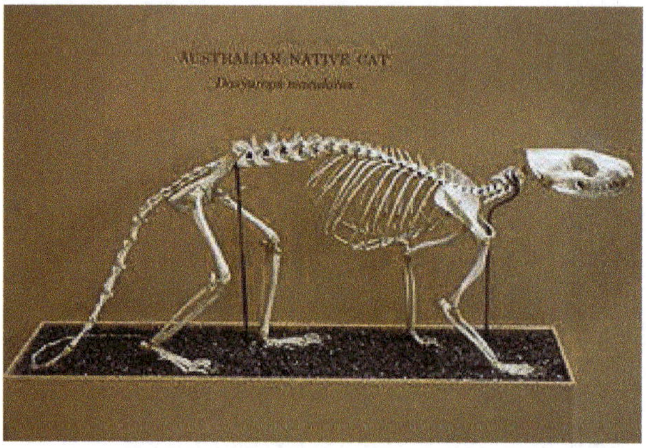

Skeleton of a tiger quoll *(Dasyurus maculatus)*.

$$y = kx^a$$

or in a logarithmic form:

$$\log y = a \log x + \log k$$

where a is the **scaling exponent** of the law. Methods for estimating this exponent from data can use type-2 regressions, such as major axis regression or reduced major axis regression, as these account for the variation in both variables, contrary to least squares regression, which does not account for error variance in the independent variable (e.g., log body mass). Other methods include measurement-error models and a particular kind of principal component analysis.

Allometry often studies shape differences in terms of ratios of the objects' dimensions. Two objects of different size, but common shape, will have their dimensions in the same ratio. Take, for example, a biological object that grows as it matures. Its size changes with age, but the shapes are similar. Studies of ontogenetic allometry often use lizards or snakes as model organisms both because they lack parental care after birth or hatching and because they exhibit a large range of body sizes between the juvenile and adult stage. Lizards often exhibit allometric changes during their ontogeny.

In addition to studies that focus on growth, allometry also examines shape variation among individuals of a given age (and sex), which is referred to as static allometry. Comparisons of species are used to examine interspecific or evolutionary allometry (see also: Phylogenetic comparative methods).

ISOMETRIC SCALING AND GEOMETRIC SIMILARITY

Isometric scaling happens when proportional relationships are preserved as size changes during growth or over evolutionary time. An example is found in frogs — aside from a brief period during the few weeks after metamorphosis, frogs grow isometrically. Therefore, a frog whose legs are as long as its body will retain that relationship throughout its life, even if the frog itself increases in size tremendously.

Scaling range for different organisms		
Group	**Factor**	**Length range**
Insects	1000	10^{-4} to 10^{-1} m
Fish	1000	10^{-2} to 10^{+1} m
Mammals	1000	10^{-1} to 10^{+2} m
Vascular plants	10,000	10^{-2} to 10^{+2} m
Algae	100,000	10^{-5} to 10^{0} m

Isometric scaling is governed by the square-cube law. An organism which doubles in length isometrically will find that the surface area available to it will increase fourfold, while its volume and mass will increase by a factor of eight. This can present problems for organisms. In the case of above, the animal now has eight times the biologically active tissue to support, but the surface area of its respiratory organs has only increased fourfold, creating a mismatch between scaling and physical demands. Similarly, the organism in the above example now has eight times the mass to support on its legs, but the strength of its bones and muscles is dependent upon their cross-sectional area, which has only increased fourfold. Therefore, this hypothetical organism would experience twice the bone and muscle loads of its smaller version. This mismatch can be avoided either by being "overbuilt" when small or by changing proportions during growth, called allometry.

Isometric scaling is often used as a null hypothesis in scaling studies, with 'deviations from isometry' considered evidence of physiological factors forcing allometric growth.

ALLOMETRIC SCALING

Allometric scaling is any change that deviates from isometry. A classic example discussed by Galileo in his *Dialogues Concerning Two New Sciences* is the skeleton of mammals.

The skeletons become much more robust and massive relative to the size of the body as the body size increases. Allometry is often expressed in terms of a scaling exponent based on body mass, or body length (Snout-vent length, total length etc.). A perfectly isometrically scaling organism would see all volume-based properties change proportionally to the body mass, all surface area-based properties change with mass to the power 2/3, and all length-based properties change with mass to the 1/3 power. If, after statistical analyses, for example, a volume-based property was found to scale to mass to the 0.9 power, then this would be called "negative allometry", as the values are smaller than predicted by isometry. Conversely, if a surface area based property scales to mass to the 0.8 power, the values are higher than predicted by isometry and the organism is said to show "positive allometry". One example of positive allometry occurs among species of monitor lizards (family Varanidae), in which the limbs are relatively longer in larger-bodied species. The same is true for some fishes, e.g. the muskellunge, whose weight grows with about the power of 3.325 of its length. A 30-inch (76 cm) muskellunge will weigh about 8 pounds (3.6 kg), while a 40-inch (100 cm) muskellunge will weigh about 18 lb (8.2 kg), as 33% longer while more than double the weight.

DETERMINING IF A SYSTEM IS SCALING WITH ALLOMETRY

To determine whether isometry or allometry is present, an expected relationship between variables needs to be determined to compare data to. This is important in determining if the scaling relationship in a dataset deviates from an expected relationship (such as those that follow isometry). The use of tools such as dimensional analysis is very helpful in determining expected slope. This 'expected' slope, as it is known, is essential for detecting allometry because scaling variables are comparisons to other things. Saying that mass scales with a slope of 5 in relation to length doesn't have much meaning unless knowing the isometric slope is 3, meaning in this case, the mass is increasing extremely fast. For example, different sized frogs should be able to jump the same distance according to the geometric similarity model proposed by Hill in 1950 and interpreted by Wilson in 2000, but in actuality larger frogs do jump longer distances. Dimensional analysis is extremely useful for balancing units in an equation or in this case, determining expected slope.

A few dimensional examples follow (M=Mass, L=Length, V=Volume, which is also L cubed because a volume is merely length cubed):

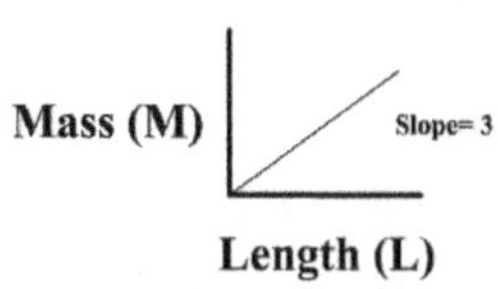

Explanation: Mass=Volume=L^3
Mass/Length= 3/1=3. Expected slope should be 3.

Allometric relations show as straight lines when plotted on double-logarithmic axes

To find the expected slope for the relationship between mass and the characteristic length of an animal (see figure), the units of mass ($M=L^3$, because mass is a volume; volumes are lengths cubed) from the Y-axis are divided by the units of the X-axis (in this case, L). The expected slope on a double-logarithmic plot of L^3/ L^1 in this case is 3 ($\log_{10}(L^3)/\log_{10}(L^1)=3$). This is the slope of a straight line, but most data gathered in science do not fall neatly in a straight line, so data transformations are useful. It is also important to keep in mind what is being compared in the data. Comparing a characteristic such as head length to head width might yield different results from comparing head length to body length. Therefore,, different characteristics may scale differently.

A common way to analyze data such as those collected in scaling is to use log-transformation. It is beneficial to transform both axes using logarithms and then perform a linear regression. This will normalize the data set and make it easier to analyze trends using the slope of the line. Before analyzing data though, it is important to have a predicted slope of the line to compare the analysis to. After data are log-transformed and linearly regressed, comparisons can then use least squares regression with 95% confidence intervals or reduced major axis analysis. Sometimes the two analyses can yield different results, but often they do not. If the expected slope is outside the confidence intervals, then there is allometry present. If mass in this imaginary animal scaled with a slope of 5 and this was a statistically significant value, then mass would scale very fast in this animal versus the expected value. It would scale with positive allometry. If the expected slope were 3 and in reality in a certain organism mass scaled with 1 (assuming this slope is statistically significant), then it would be negatively allometric.

Another example: Force is dependent on the cross-sectional area of muscle (CSA), which is L^2. If comparing force to a length, then the expected slope is 2. Alternatively, this analysis may be accomplished with a power regression. Plot the relationship between the data onto a graph. Fit this to a power curve (depending on the stats program, this can be done multiple ways), and it will give an equation with the form: $y=Zx^n$, where n is the number. That "number" is the relationship between the data points. The downside, to this form of analysis, is that it makes it a little more difficult to do statistical analyses.

PHYSIOLOGICAL SCALING

Many physiological and biochemical processes (such as heart rate, respiration rate or the maximum reproduction rate) show scaling, mostly associated with the ratio between surface area and mass (or volume) of the animal. The metabolic rate of an individual animal is also subject to scaling.

Metabolic Rate and Body Mass

In plotting an animal's basal metabolic rate (BMR) against the animal's own body mass, a logarithmic straight line is obtained, indicating a power-law dependence. Overall metabolic rate in animals is generally accepted to show negative allometry, scaling to mass to a power ≈ 0.75, known as Kleiber's law, 1932. This means that larger-bodied species (e.g., elephants) have lower mass-specific metabolic rates and lower heart rates, as compared with smaller-bodied species (e.g., mice). The straight line generated from a double logarithmic scale of metabolic rate in relation to body mass is known as the mouse-to-elephant curve. These relationships of metabolic rates, times, and internal structure have been explained as, an elephant is approximately a blown-up gorilla, which is itself a blown-up mouse.

Max Kleiber contributed the following allometric equation for relating the BMR to the body mass of an animal. Statistical analysis of the intercept did not vary from 70 and the slope was not varied from 0.75, thus:

$$\text{Metabolic rate} = 70M^{0.75}$$

where M is body mass, and metabolic rate is measured in kcal per day.

Consequently, the body mass itself can explain the majority of the variation in the BMR. After the body mass effect, the taxonomy of the animal plays the next most significant role in the scaling of the BMR. The speculation that environmental conditions play a role in BMR can only be properly investigated once the role of taxonomy is established. The challenge with this lies in the fact that a shared environment also indicates a common evolutionary history and thus a close taxonomic relationship. There are strides currently in research to overcome these hurdles. For example, an analysis in muroid rodents, the mouse, hamster, and vole type, took into account taxonomy. Results revealed the hamster (warm dry habitat) had lowest BMR and the mouse (warm wet dense habitat) had the highest BMR. Larger organs could explain the high BMR groups, along with their higher daily energy needs. Analyses such as these demonstrate the physiological adaptations to environmental changes that animals undergo.

Energy metabolism is subjected to the scaling of an animal and can be overcome by an individual's body design. The metabolic scope for an animal is the ratio of resting and maximum rate of metabolism for that particular species as determined by oxygen consumption. Oxygen consumption V_{O2} and maximum oxygen consumption VO2 max. Oxygen consumption in species that differ in body size and organ system dimensions show a similarity in their charted V_{O2} distributions indicating that, despite the complexity of their systems, there is a power law dependence of similarity; therefore, universal patterns are observed in diverse animal taxonomy.

Across a broad range of species, allometric relations are not necessarily linear on a log-log scale. For example, the maximal running speeds of mammals show a complicated relationship with body mass, and the fastest sprinters are of intermediate body size.

Allometric Muscle Characteristics

The muscle characteristics of animals are similar in a wide range of animal sizes, though muscle sizes and shapes can and often do vary depending on environmental constraints placed on them. The muscle tissue itself maintains its contractile characteristics and does not vary depending on the size of the animal. Physiological scaling in muscles affects the number of muscle fibers and their intrinsic speed to determine the maximum power and efficiency of movement in a given animal. The speed of muscle recruitment varies roughly in inverse proportion to the cube root of the animal's weight, such as the intrinsic frequency of the sparrow's flight muscle compared to that of a stork's.

$$\text{freq} = \frac{1}{\text{mass}^{1/3}}$$

For inter-species allometric relations related to such ecological variables as maximal reproduction rate, attempts have been made to explain scaling within the context of dynamic energy budget theory and the metabolic theory of ecology. However, such ideas have been less successful.

Allometry of Legged Locomotion

Methods of study

Allometry has been used to study patterns in locomotive principles across a broad range of species. Such research has been done in pursuit of a better understanding of animal locomotion, including the factors that different gaits seek to optimize. Allometric trends observed in extant animals have even been combined with evolutionary algorithms to form realistic hypotheses concerning the locomotive patterns of extinct species. These studies have been made possible by the remarkable similarities among disparate species' locomotive kinematics and dynamics, "despite differences in morphology and size".

Allometric study of locomotion involves the analysis of the relative sizes, masses, and limb structures of similarly shaped animals and how these features affect their movements at different speeds. Patterns are identified based on dimensionless Froude numbers, which incorporate measures of animals' leg lengths, speed or stride frequency, and weight.

Alexander incorporates Froude-number analysis into his "dynamic similarity hypothesis" of gait patterns. Dynamically similar gaits are those between which there are constant coefficients that can relate linear dimensions, time intervals, and forces. In other words, given a mathematical description of gait A and these three coefficients, one could produce gait B, and vice versa. The hypothesis itself is as follows: "animals of different sizes tend to move in dynamically similar fashion whenever the ratio of their speed allows it." While the dynamic similarity hypothesis may not be a truly unifying principle of animal gait patterns, it is a remarkably accurate heuristic.

It has also been shown that living organisms of all shapes and sizes utilize spring mechanisms in their locomotive systems, probably in order to minimize the energy cost of locomotion. The allometric study of these systems has fostered a better understanding of why spring mechanisms are so common, how limb compliance varies with body size and speed, and how these mechanisms affect general limb kinematics and dynamics.

Principles of Legged Locomotion Identified through Allometry

- Alexander found that animals of different sizes and masses traveling with the same Froude number consistently exhibit similar gait patterns.

- Duty factors—percentages of a stride during which a foot maintains contact with the ground—remain relatively constant for different animals moving with the same Froude number.
- The dynamic similarity hypothesis states that animals of different sizes tend to move in dynamically similar fashion whenever the ratio of their speed allows it.
- Body mass has even more of an effect than speed on limb dynamics.
- Leg stiffness, $k_{leg} = \frac{\text{peak force}}{\text{peak displacement}}$, is proportional to $M^{0.67}$, where M is body mass.
- Peak force experienced throughout a stride is proportional to $M^{0.97}$.
- The amount by which a leg shortens during a stride (i.e. its peak displacement) is proportional to $M^{0.30}$.
- The angle swept by a leg during a stride is proportional to $M^{-0.034}$.
- The mass-specific work rate of a limb is proportional to $M^{0.11}$.

Drug Dose Scaling

The physiological effect of drugs and other substances in many cases scales allometrically.

West and Brown in 2005 derived a hydrodynamic theory to explain the universal fact that metabolic rate scales as the ¾ power with body weight. They also showed why lifespan scales as the +¼ power, heart rate as the -¼ power, and hence why all species have a similar number, about 1.5 billion, heartbeats during their lifespan. Blood flow (+¾) and resistance (-¾) scale in the same way, leading to blood pressure being constant across species.

Hu and Hayton in 2001 discussed whether the basal metabolic rate scale is a ⅔ or ¾ power of body mass. The exponent of ¾ might be used for substances that are eliminated mainly by metabolism, or by metabolism and excretion combined, while ⅔ might apply for drugs that are eliminated mainly by renal excretion.

An online allometric scaler of drug doses based on the above work is available.

The US Food and Drug Administration (FDA) published guidance in 2005 giving a flow chart that presents the decisions and calculations used to generate the maximum recommended starting dose in drug clinical trials from animal data.

ALLOMETRIC SCALING IN FLUID LOCOMOTION

The mass and density of an organism have a large effect on the organism's locomotion through a fluid. For example, a tiny organisms uses flagella and can effectively move through a fluid it is suspended in. Then on the other scale a blue whale that is much more massive and dense in comparison with the viscosity of the fluid, compared to a bacterium in the same medium. The way in which the fluid interacts with the external boundaries of the organism is important with locomotion through the fluid. For streamlined swimmers the resistance or drag determines the performance of the organism. This drag or resistance can be seen in two distinct flow patterns. There is Laminar Flow where the fluid is relatively uninterrupted after the organism moves through it. Turbulent flow is the opposite, where the fluid moves roughly around an organisms that creates vortices that absorb energy from the propulsion or momentum of the organism. Scaling also affects locomotion through a fluid because of the energy needed to propel an organism and to keep up velocity through momentum. The rate of oxygen consumption per gram body size decreases consistently with increasing body size. (Knut Schmidt-Nielson 2004)

In general, smaller, more streamlined organisms create laminar flow ($R < 0.5x106$), whereas larger, less streamlined organisms produce turbulent flow ($R > 2.0\times106$). Also, increase in velocity (V) increases turbulence, which can be proved using the Reynolds equation. In nature however, organisms such as a 6'-6" dolphin moving at 15 knots does not have the appropriate Reynolds numbers for laminar flow $R = 107$, but exhibit it in nature. Mr. G.A Steven observed and documented dolphins moving at 15 knots alongside his ship leaving a single trail of light when phosphorescent activity in the sea was high. The factors that contribute are:

- Surface area of the organism and its effect on the fluid in which the organism lives is very important in determining the parameters of locomotion.
- The Velocity of an organism through fluid changes the dynamic of the flow around that organism and as velocity increases the shape of the organism becomes more important for laminar flow.
- Density and viscosity of fluid.
- Length of the organism is factored into the equation because the surface area of just the front 2/3 of the organism has an effect on the drag

The resistance to the motion of an approximately stream-lined solid through a fluid can be expressed by the formula: $C_{f\rho}$(total surface)$V2/2$

V = velocity

ρ = density of fluid

$C_f = 1.33R - 1$ (laminar flow) R = Reynolds number

Reynolds number $[R] = VL/v$

V = velocity

L = axial length of organism

v = kinematic viscosity (viscosity/density)

Notable Reynolds numbers:

$R < 0.5x106$ = laminar flow threshold

$R > 2.0x106$ = turbulent flow threshold

Scaling also has an effect on the performance of organisms in fluid. This is extremely important for marine mammals and other marine organisms that rely on atmospheric oxygen to survive and carry out respiration. This can affect how fast an organism can propel itself efficiently and more importantly how long it can dive, or how long and how deep an organism can stay underwater. Heart mass and lung volume are important in determining how scaling can affect metabolic function and efficiency. Aquatic mammals, like other mammals, have the same size heart proportional to their bodies.

Mammals have a heart that is about 0.6% of the total body mass across the board from a small mouse to a large Blue Whale. It can be expressed as: Heart Weight = 0.006Mb1.0, where Mb is the body mass of the individual. Lung volume is also directly related to body mass in mammals (slope = 1.02). The lung has a volume of 63 ml for every kg of body mass. In addition, the tidal volume at rest in an individual is 1/10 the lung volume. Also respiration costs with respect to oxygen consumption is scaled in the order of Mb.75. This shows that mammals, regardless of size, have the same size respiratory and cardiovascular systems and it turn have the same amount of blood: About 5.5% of body mass. This means that for a similarly designed marine mammals, the larger the individual the more efficiently they can travel compared to a smaller individual. It takes the same effort to move one body length whether the individual is one meter or ten meters. This can explain why large whales can migrate far distance in the oceans and not stop for rest. It is metabolically less expensive to be larger in body size. This goes for terrestrial and flying animals as well. In fact, for an organism to move any distance, regardless of type from elephants to centipedes, smaller animals consume more oxygen per unit body mass than larger ones. This metabolic advantage that larger animals have makes it possible for larger marine mammals to dive for longer durations of time than their smaller counterparts. That the heart rate is lower means that larger animals can carry more blood, which carries more oxygen. Then in conjuncture with the fact that mammals reparation costs scales in the order of Mb.75 shows how an advantage can be had in having a larger body mass. More simply, a larger whale can hold more oxygen and at the same time demand less metabolically than a smaller whale.

Traveling long distances and deep dives are a combination of good stamina and also moving an efficient speed and in an efficient way to create laminar flow, reducing drag and turbulence. In sea water as the fluid, it traveling long distances in large mammals, such as whales, is facilitated by their neutral buoyancy and have their mass completely supported by the density of the sea water. On land, animals have to expend a portion of their energy during locomotion to fight the effects of gravity.

Flying organisms such as birds are also considered moving through a fluid. In scaling birds of similar shape, it has also been seen that larger individuals have less metabolic cost per kg than smaller species, which would be expected because it holds true for every other form of animal. Birds also have a variance in wing beat frequency. Even with the compensation of larger wings per unit body mass, larger birds also have a slower wing beat frequency, which allows larger birds to fly at higher altitudes, longer distances, and faster absolute speeds than smaller birds. Because of the dynamics of lift-based locomotion and the fluid dynamics, birds have a U-shaped curve for metabolic cost and velocity. Because flight, in air as the fluid, is metabolically more costly at the lowest and the highest velocities. On the other end, small organisms such as insects can make gain advantage from the viscosity of the fluid (air) that they are moving in. A wing-beat timed perfectly can effectively uptake energy from the previous stroke. (Dickinson 2000) This form of wake capture allows an organism to recycle energy from the fluid or vortices within that fluid created by the organism itself. This same sort of wake capture occurs in aquatic organisms as well, and for organisms of all sizes. This dynamic of fluid locomotion allows smaller organisms to gain advantage because the effect on them from the fluid is much greater because of their relatively smaller size.

ALLOMETRIC ENGINEERING

Allometric engineering is a method for manipulating allometric relationships within or among groups.

ALLOMETRY IN CHARACTERISTICS OF A CITY

Arguing that there are a number of analogous concepts and mechanisms between cities and biological entities, Bettencourt et al. showed a number of scaling relationships between observable properties of a city and the city size. GDP, "supercreative" employment, number of inventors, crime, spread of disease, and even pedestrian walking speeds scale with city population.

EXAMPLES

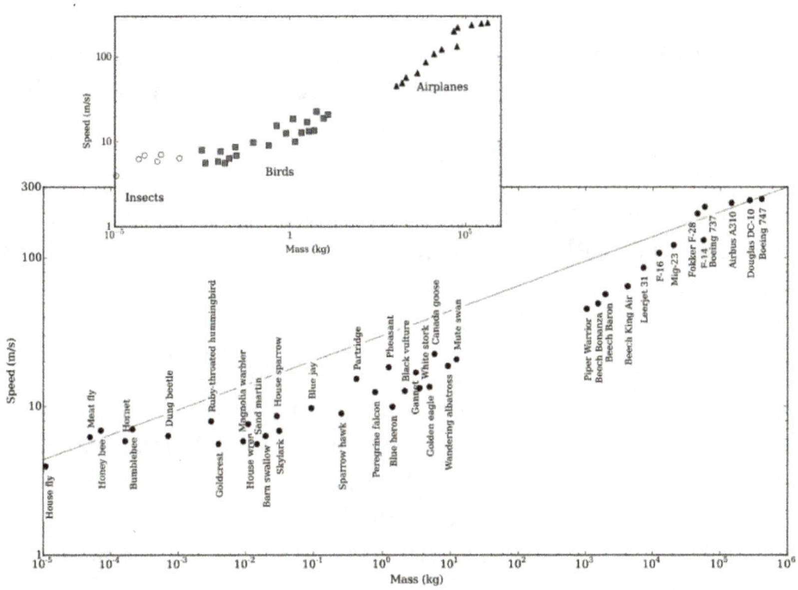

Allometric law of cruising speed versus body mass

Some examples of allometric laws:

- Kleiber's law, metabolic rate q_0 is proportional to body mass M raised to the $3/4$ power:

$$q_0 \sim M^{\frac{3}{4}}$$

- breathing and heart rate t are both inversely proportional to body mass M raised to the $1/4$ power:

$$t \sim M^{-\frac{1}{4}}$$

- mass transfer contact area A and body mass M:

$$A \sim M^{\frac{7}{8}}$$

- the proportionality between the optimal cruising speed V_{opt} of flying bodies (insects, birds, airplanes) and body mass M raised to the power $1/6$:

$$V_{\text{opt}} \sim M^{\frac{1}{6}}$$

DETERMINANTS OF SIZE IN DIFFERENT SPECIES

Many factors go into the determination of body mass and size for a given animal. These factors often affect body size on an evolutionary scale, but conditions such as availability of food and habitat size can act much more quickly on a species. Other examples include the following:

- Physiological design

Basic physiological design plays a role in the size of a given species. For example, animals with a closed circulatory system are larger than animals with open or no circulatory systems.

- Mechanical design

Mechanical design can also determine the maximum allowable size for a species. Animals with tubular endoskeletons tend to be larger than animals with exoskeletons or hydrostatic skeletons.

- Habitat

An animal's habitat throughout its evolution is one of the largest determining factors in its size. On land, there is a positive correlation between body mass of the top species in the area and available land area. However, there are a much greater number of "small" species in any given area. This is most likely determined by ecological conditions, evolutionary factors, and the availability of food; a small population of large predators depend on a much greater population of small prey to survive. In an aquatic environment, the largest animals can grow to have a much greater body mass than land animals where gravitational weight constraints are a factor.

ALTERNATIVE STABLE STATE

In ecology, the theory of **alternative stable states** (sometimes termed **alternate stable states** or **alternative stable equilibria**) predicts that ecosystems can exist under multiple "states" (sets of unique biotic and abiotic conditions). These alternative states are non-transitory and therefore considered stable over ecologically-relevant timescales. Ecosystems may transition from one stable state to another, in what is known as a state shift (sometimes termed a phase shift or regime shift), when perturbed. Due to ecological feedbacks, ecosystems display resistance to state shifts and therefore tend to remain in one state unless perturbations are large enough. Multiple states may persist under equal environmental conditions, a phenomenon known as hysteresis. **Alternative stable state** theory suggests that discrete states are separated by ecological thresholds, in contrast to ecosystems which change smoothly and continuously along an environmental gradient.

Theory

Alternative stable state theory was first proposed by Richard Lewontin (1969), but other early key authors include Holling (1973), Sutherland (1974), May (1977), and Scheffer et al. (2001). In the broadest sense, alternative stable state theory proposes that a change in ecosystem conditions can result in an abrupt shift in the state of the ecosystem, such as a change in population or community composition. Ecosystems can persist in states that are considered stable (i.e., can exist for relatively long periods of time). Intermediate states are considered unstable and are, therefore, transitory. Because ecosystems are resistant to state shifts, significant perturbations are usually required to overcome ecological thresholds and cause shifts from one stable state to another. The resistance to state shifts is known as "resilience" (Holling 1973).

State shifts are often illustrated heuristically by the ball-in-cup model (Holling, C.S. et al. (1995) Biodiversity in the functioning of ecosystems: an ecological synthesis. In Biodiversity Loss, Ecological and Economical Issues (Perrings, C.A. et al., eds), pp. 44–83, Cambridge University Press). A ball, representing the ecosystem, exists on a surface where any point along the surface represents a possible state. In the simplest model, the landscape consists of two valleys separated by a hill. When the ball is in a valley, or a "domain of attraction," it exists in a stable state and must be perturbed to move from this state. In the absence of perturbations, the ball will always roll downhill and therefore will tend to stay in the valley (or stable state). State shifts can be viewed from two different viewpoints, the "community perspective" and the "ecosystem perspective." The ball can only move between stable states in two ways: (1) moving the ball or (2) altering the landscape. The community perspective is analogous to moving the ball, while the ecosystem perspective is analogous to altering the landscape.

These two viewpoints consider the same phenomenon with different mechanisms. The community perspective considers ecosystem variables (which change relatively quickly and are subject to feedbacks from the system), whereas the ecosystem perspective considers ecosystem parameters (which change relatively slowly and operate independent of the system). The community context considers a relatively constant environment in which multiple stable states are accessible to populations or communities. This definition is an extension of stability analysis of populations (e.g., Lewontin 1969; Sutherland 1973) and communities (e.g., Drake 1991; Law and Morton 1993). The ecosystem context focuses on the effect of exogenic "drivers" on communities or ecosystems (e.g., May 1977; Scheffer et al. 2001; Dent et al. 2002). Both definitions are explored within this article.

Ecological Community

Ecosystems can shift from one state to another via a significant perturbation directly to state variables. State variables are quantities that change quickly (in ecologically-relevant time scales) in response to feedbacks from the system (i.e., they are dependent on system feedbacks), such as population densities. This perspective requires that different states can exist simultaneously under equal environmental conditions, since the ball moves only in response to a state variable change.

For example, consider a very simple system with three microbial species. It may be possible for the system to exist under different community structure regimes depending on initial conditions (e.g., population densities or spatial arrangement of individuals) (Kerr et al. 2002). Perhaps under certain initial densities or spatial configurations, one species dominates over all others, while under different initial conditions all species can mutually coexist. Because the different species interact, changes in populations affect one another synergistically to determine community structure. Under both states the environmental conditions are identical. Because the states have resilience, following small perturbations (e.g., changes to population size) the community returns to the same configuration while large perturbations may induce a shift to another configuration.

The community perspective requires the existence of alternative stable states (i.e., more than one valley) before the perturbation, since the landscape is not changing. Because communities have some level of resistance to change, they will stay in their domain of attraction (or stable state) until the perturbation is large enough to force the system into another state. In the ball-and-cup model, this would be the energy required to push the ball up and over a hill, where it would fall downhill into a different valley.

Ecosystem

It is also possible to cause state shifts in another context, by indirectly affecting state variables. This is known as the ecosystem perspective. This perspective requires a change in environmental parameters that affect the behavior of state variables. For example, birth rate, death rate, migration, and density-dependent predation indirectly alter the ecosystem state by changing population density (a state variable). Ecosystem parameters are quantities that are unresponsive (or respond very slowly) to feedbacks from the system (i.e., they are independent of system feedbacks). The stable state landscape is changed by environmental drivers, which may result in a change in the quantity of stable states and the relationship between states.

By the ecosystem perspective, the landscape of the ecological states is changed, which forces a change in the ecosystem state. Changing the landscape can modify the number, location, and resilience of stable states, as well as the unstable intermediate states. By this view, the topography in the ball-and-cup model is not static, as it is in the community perspective. This is a fundamental difference between the two perspectives.

However, it should be noted that although the mechanisms of community and ecosystem perspectives are different, the empirical evidence required for documentation of alternative stable states is the same. In addition, state shifts are often a combination of internal processes and external forces (Scheffer et al. 2001). For example, consider a stream-fed lake in which the pristine state is dominated by benthic vegetation. When upstream construction releases soils into the stream, the system becomes turbid. As a result, benthic vegetation cannot receive light and decline, increasing nutrient availability and allowing phytoplankton to dominate. In this state shift scenario the state variables changing are the populations of benthic vegetation and phytoplankton, and the ecosystem parameters are turbidity and nutrient levels. So, whether identifying mechanisms of variables or parameters is a matter of formulation (Beisner et al. 2003).

Hysteresis Effects

Hysteresis is an important concept in alternative stable state theory. In this ecological context, hysteresis refers to the existence of different stable states under the same variables or parameters. Hysteresis can be explained by "path-dependency," in which the equilibrium point for the trajectory of "A → B" is different from for "B → A." In other words, it matters which way the ball is moving across the landscape. Some ecologists (e.g., Scheffer et al. 2001) argue that hysteresis is a prerequisite for the existence of alternative stable states. Others (e.g., Beisner et al. 2003) claim that this is not so; although shifts often involve hysteresis, a system can show alternative stable states yet have equal paths for "A → B" and "B → A."

Hysteresis can occur via changes to variables or parameters. When variables are changed the ball is pushed from one domain of attraction to another, yet the same push from the other direction cannot return the ball to the original domain of attraction. When parameters are changed a modification to the landscape results in a state shift, but reversing the modification does not result in a reciprocal shift.

A real-world example of hysteresis is helpful to illustrate the concept. Coral reef systems can dramatically shift from pristine coral-dominated systems to degraded algae-dominated systems when populations grazing on algae decline. The 1983 crash of urchin populations in Caribbean reef systems released algae from top-down (herbivory) control, allowing them to overgrow corals and resulting in a shift to a degraded state. When urchins rebounded, the high (pre-crash) coral cover levels did not return, indicating hysteresis (Mumby et al. 2007).

In some cases, state shifts under hysteresis may be irreversible. For example, tropical cloud forests require high moisture levels, provided by clouds that are intercepted by the canopy (via condensation). When deforested, moisture delivery ceases. Therefore, reforestation is often unsuccessful because conditions are too dry to allow the trees to grow (Wilson and Agnew 1992).

Resilience

By their very nature, basins of attraction display resilience. Ecosystems are resistant to state shifts – they will only undergo shifts under substantial perturbations – but some states are more resilient than others. In the ball-and-cup model, a valley with steep sides has greater resilience than a shallow valley, since it would take more force to push the ball up the hill and out of the valley.

Resilience can change in stable states when environmental parameters are shifted. Often, humans influence stable states by reducing the resilience of basins of attraction. There are at least three ways in which anthropogenic forces reduce resilience (Folke et al. 2004): (1) Decreasing diversity and functional groups, often by top-down effects (e.g., overfishing); (2) altering the physico-chemical environment (e.g., climate change, pollution, fertilization); or (3) modifying disturbance regimes to which organisms are adapted (e.g., bottom trawling, coral mining, etc.). When the resilience is decreased, ecosystems can be pushed into alternative, and often less-desirable, stable states with only minor perturbations. When hysteresis effects are present, the return to a more-desirable state is sometimes impossible or impractical (given management constraints). Shifts to less-desirable states often entail a loss of ecosystem service and function, and have been documented in an array of terrestrial, marine, and freshwater environments (reviewed in Folke et al. 2004).

Evidence

State shifts via the community perspective have been induced experimentally by the addition or removal of predators, such as in Paine's (1966) work on keystone predators (i.e., predators with disproportionate influence on community structure) in the intertidal zone (although this claim is refuted by Schröder et al. 2005). Also, Beisner et al. (2003) suggest that commercially exploited fish populations can be forced between alternative stable states by fishing pressure due to Allee effects that work at very low population sizes. Once a fish population falls below a certain threshold, it will inevitably go extinct when low population densities make replacement of adults impossible due to, for example, the inability to find mates or density-dependent mortality. Since populations cannot return from extinction, this is an example of an irreversible state shift.

Most work on alternative stable states has been theoretical, using mathematical models and simulations to test ecological hypotheses. Other work has been conducted using empirical evidence from surveying, historical records, or comparisons across spatial scales. There has been a lack of direct, manipulative experimental tests for alternative stable states. This is especially true for studies outside of controlled laboratory conditions, where state shifts have been documented for cultures of microorganisms. Empirical evidence for the existence of alternative stable states is vital to advancing the idea beyond theory. Schröder et al. (2005) reviewed the current ecological literature for alternative stable states and found 35 direct experiments, of which only 21 were deemed valid. Of these, 62% (14) showed evidence for and 38% (8) showed no evidence for alternative stable states. However, the Schröder et al. (2005) analysis required evidence of hysteresis, which is not necessarily a prerequisite for alternative stable states. Other authors (e.g., Scheffer et al. 2001; Folke et al. 2004) have had less-stringent requirements for the documentation of alternative stable states.

Although alternative stable state theory is still in its infancy, empirical evidence has been collected from a variety of biomes. The Sahara region, once vegetated, suddenly collapsed into a desert about 5,000 to 6,000 years ago due to climate change (Scheffer and Carpenter 2003). Recent alteration of disturbance states via fire suppression has caused state shifts in fire-adapted ecosystems of the southeastern United States (Peterson 2002).

In the northeastern United States alternative stable states have been experimentally documented in hardwood forest food webs (Schmitz 2004). Spiders exhibit top-down control on generalist herbivore grasshoppers, causing them to hide in and forage on a competitively dominant plant species. Herbivory on the superior plant competitor reduces dominance, thereby increasing plant diversity. Under experimental removal of spiders, grasshoppers were released from predation and consumed plants without preference for the competitively-superior species. This top competitor outcompeted other species, resulting in lower overall plant diversity. Furthermore, reintroductions of spiders failed to return the system to high plant diversity, indicating ecological thresholds. Ponds, lakes, and other small bodies of water are ideal for studying alternative stable states because they are relatively self-contained (Holling 1973). There has been great interest in state shifts in shallow temperate lakes between clear-water and turbid-water states. Nutrient loading can shift clear lakes to turbid systems by causing phytoplankton blooms (Scheffer 1997). Feedbacks maintaining the turbid state include shading out benthic macrophytes, which otherwise stabilize sediments and reduce rapid nutrient cycling. Hysteresis may occur if reductions in nutrient input do not return the clear-water state. Alternative stable states have also been shown for freshwater faunal communities. Initial conditions, such as densities of snails, can influence the movement of an ecosystem into one of several alternative stable equilibria (Chase 2003). Additionally, alternative stable states have been documented experimentally in coastal marine ecosystems. Arctic salt marshes of Hudson Bay, Manitoba have been shown to shift from a vegetated state to an unvegetated state due to overgrazing by snow geese (Handa et al. 2002). When geese were experimentally removed, recovery of the system only occurred under ideal soil conditions, indicating hysteresis.

Verifying the existence of alternative stable states carries profound implications for ecosystem management. If stable states exist, gradual changes in environmental factors may have little effect on a system until a threshold is reached, at which point a catastrophic state shift may occur. Understanding the nature of these thresholds will help inform the design of monitoring programs, ecosystem restoration, and other management decisions. Managers are particularly interested in the potential of hysteresis, since it may be difficult to recover from a state shift (Beisner et al. 2003). The mechanisms of feedback loops that maintain stable states are important to understand if we hope to effectively manage an ecosystem with alternative stable states.

REFERENCES

1. Small, Christopher G. (1996). The Statistical Theory of Shape. Springer. p. 4. ISBN 0-387-94729-9.
2. Damuth J (February 2001). "Scaling of growth: plants and animals are not so different". Proc. Natl. Acad. Sci. U.S.A. 98 (5): 2113–4. Bibcode:2001PNAS...98.2113D. doi:10.1073/pnas.051011198. PMC 33381 . PMID 11226197.

3. Otto Snell (1892). "Die Abhängigkeit des Hirngewichts von dem Körpergewicht und den geistigen Fähigkeiten". Arch. Psychiatr. 23 (2): 436–446. doi:10.1007/BF01843462.
4. Thompson, D'Arcy W (1992). On Growth and Form(Canto ed.). Cambridge University Press. ISBN 978-0-521-43776-9.
5. Huxley, Julian S. (1972). Problems of Relative Growth (2nd ed.). New York: Dover. ISBN 0-486-61114-0.
6. Garland, T., Jr.; P. L. Else (March 1987). "Seasonal, sexual, and individual variation in endurance and activity metabolism in lizards" (PDF). Am J Physiol. 252 (3 Pt 2): R439–49. PMID 3826408.
7. Vogel, Steven (1988). Life's Devices : The Physical World of Animals and Plants. Princeton University Press. p. 39. ISBN 978-0-691-02418-9. Retrieved 29 March 2014.
8. Emerson SB (September 1978). "Allometry and Jumping in Frogs: Helping the Twain to Meet". Evolution. 32 (3): 551–564. doi:10.2307/2407721. JSTOR 2407721.
9. Schmidt-Nielsen 1984
10. Christian, A.; Garland T., Jr. (1996). "Scaling of limb proportions in monitor lizards (Squamata: Varanidae)"(PDF). Journal of Herpetology. 30 (2): 219–230. doi:10.2307/1565513. JSTOR 1565513.
11. R. O. Anderson and R. M. Neumann, Length, Weight, and Associated Structural Indices, in Fisheries Techniques, second edition, B.E. Murphy and D.W. Willis, eds., American Fisheries Society, 1996.
12. Pennycuick, Colin J. (1992). Newton Rules Biology. Oxford University Press. p. 111. ISBN 0-19-854021-3.
13. Schmidt-Nielsen 1984, p. 237
14. Gibbings, J.C. (2011). Dimensional Analysis. Springer. ISBN 978-1-84996-317-6.
15. Hill, A.V. (April 1950). "The dimensions of animals and their muscular dynamics". Science Progress. 150 (4176): 820. Bibcode:1949Natur.164R.820.. doi:10.1038/164820b0.
16. Wilson RS, Franklin CE, James RS (June 2000). "Allometric scaling relationships of jumping performance in the striped marsh frog Limnodynastes peronii". J. Exp. Biol. 203 (Pt 12): 1937–46. PMID 10821750.
17. Robinson, Michael; Motta, Philip (2002). "Patterns of growth and the effects of scale on the feeding kinematics of the nurse shark (Ginglymostoma cirratum)". Journal of Zoology, London. 256 (4): 449–462. doi:10.1017/S0952836902000493.
18. O'Hara, R.B.; Kotze, D.J. (2010). "Do not log-transform count data". Methods in Ecology and Evolution. 1 (2): 118–122. doi:10.1111/j.2041-210X.2010.00021.x.
19. Willmer, Pat (2009). Environmental Physiology of Animals. Wiley-Blackwell.
20. Bettencourt LM, Lobo J, Helbing D, Kühnert C, West GB (April 2007). "Growth, innovation, scaling, and the pace of life in cities". Proc. Natl. Acad. Sci. U.S.A.

104 (17): 7301–6. Bibcode:2007PNAS..104.7301B. doi:10.1073/pnas.0610172104. PMC 1852329 . PMID 17438298.

21. Labra FA, Marquet PA, Bozinovic F (June 2007). "Scaling metabolic rate fluctuations". Proc. Natl. Acad. Sci. U.S.A. 104 (26): 10900–3. Bibcode:2007PNAS..10410900L. doi:10.1073/pnas.0704108104. PMC 1904129 . PMID 17578913.
22. Garland Jr., T. (1983). "The relation between maximal running speed and body mass in terrestrial mammals"(PDF). Journal of Zoology, London. 199 (2): 157–170. doi:10.1111/j.1469-7998.1983.tb02087.x.
23. Chappell, R. (1989). "Fitting bent lines to data, with applications to allometry". Journal of Theoretical Biology. 138 (2): 235–256. doi:10.1016/S0022-5193(89)80141-9. PMID 2607772.
24. Daley, Monica A.; Usherwood, James R. (2010). "Two explanations for the compliant running paradox: reduced work of bouncing viscera and increased stability in uneven terrain". Biology Letters. 6 (3): 418–421. doi:10.1098/rsbl.2010.0175. PMC 2880072 . PMID 20335198.
25. Farley CT, Glasheen J, McMahon TA (December 1993). "Running springs: speed and animal size". J. Exp. Biol. 185 (1): 71–86. PMID 8294853.
26. ^ Jump up to:a b c Sellers, William Irving; Manning, Phillip Lars (2007). "Estimating dinosaur maximum running speeds using evolutionary robotics". Proceedings of the Royal Society B. 274 (1626): 2711–6. doi:10.1098/rspb.2007.0846. PMC 2279215 . PMID 17711833.
27. ^ Jump up to:a b c d e f g h Alexander, R. McN. (1984). "The gaits of bipedal and quadrupedal animals". The International Journal of Robotics Research. 3 (2): 49–59. doi:10.1177/027836498400300205.
28. Roberts, Thomas J.; Azizi, Emanuel (2011). "Flexible mechanisms: the diverse roles of biological springs in vertebrate movement". The Journal of Experimental Biology. 214 (3): 353–361. doi:10.1242/jeb.038588.
29. West, G. B. (2005). "The origin of allometric scaling laws in biology from genomes to ecosystems: towards a quantitative unifying theory of biological structure and organization". Journal of Experimental Biology. 208 (9): 1575–1592. doi:10.1242/jeb.01589. ISSN 0022-0949.
30. T. M. Hu; W. L. Hayton (2001). "Allometric scaling of xenobiotic clearance: uncertainty versus universality". AAPS PharmSci. 3 (4): E29. doi:10.1208/ps030429. PMC 2751218 . PMID 12049492.
31. "Allometric Scaling Calculator". Clymer.altervista.org. 13 May 2012. Retrieved 15 December 2015. Online allometric scaling calculator, with explanation and source.
32. US FDA: Estimating the Safe Starting Dose in Clinical Trials for Therapeutics in Adult Healthy Volunteers, July 2005
33. Schmidt-Nielsen, Knut (10 April 1997). Animal Physiology: Adaptation and Environment (5th ed.). Cambridge University Press. ISBN 978-0-521-57098-5.

34. Dickinson MH, Farley CT, Full RJ, Koehl MA, Kram R, Lehman S (April 2000). "How animals move: an integrative view". Science. 288 (5463): 100–6. Bibcode:2000Sci...288..100D. doi:10.1126/science.288.5463.100. PMID 10753108.

35. Sinervo, B.; Huey, R. (1990). "Allometric Engineering: An Experimental Test of the Causes of Interpopulational Differences in Performance" (PDF). Science. 248 (4959): 1106–9. Bibcode:1990Sci...248.1106S. doi:10.1126/science.248.4959.1106.

36. Bornstein MH, Bornstein HG (19 February 1976). "The Pace of Life". Nature. 259 (5544): 557–9. Bibcode:1976Natur.259..557B. doi:10.1038/259557a0.

37. Burness, G. P.; Diamond, Jared; Flannery, Timothy (2001). "Dinosaurs, dragons, and dwarfs: The evolution of maximal body size". Proc. Natl. Acad. Sci. U.S.A. 98 (25): 14518–23. Bibcode:2001PNAS...9814518B. doi:10.1073/pnas.251548698. PMC 64714 . PMID 11724953.

38. Beisner, B. E.; Haydon, D. T.; Cuddington, K. (2003). "Alternative stable states in ecology". Frontiers in Ecology and the Environment. 1 (7): 376–382. doi:10.1890/1540-9295(2003)001[0376:ASSIE]2.0.CO;2.

39. Chase, J. M. (2003). "Experimental evidence for alternative stable state equilibria in a benthic pond food web". Ecology Letters. 6 (8): 733–741. doi:10.1046/j.1461-0248.2003.00482.x.

40. Dent, C. L.; Cummings, G. S.; Carpenter, S. R. (2002). "Multiple states in river and lake ecosystems". Philosophical Transactions of the Royal Society of London Series B – Biological Sciences. 357 (1421): 635–645. doi:10.1098/rstb.2001.0991.

41. Drake, J. A. (1991). "Community-assembly mechanics and the structure of an experimental species ensemble". The American Naturalist. 137 (1): 1–26. doi:10.1086/285143. JSTOR 2462154.

42. Folke, C.; Carpenter, S.; Walker, B.; Scheffer, M.; Elmqvist, T.; Gunderson, L.; Holling, C. S. (2004). "Regime shifts, resilience, and biodiversity in ecosystem management". Annual Review of Ecology, Evolution, and Systematics. 35: 557–581. doi:10.1146/annurev.ecolsys.35.021103.105711. JSTOR 30034127.

43. Handa, I. T.; Harmsen, R.; Jefferies, R. L. (2002). "Patterns of Vegetation Change and the Recovery Potential of Degraded Areas in a Coastal Marsh System of the Hudson Bay Lowlands". Journal of Ecology. 90 (1): 86–99. doi:10.1046/j.0022-0477.2001.00635.x.

44. Holling, C. S. (1973). "Resilience and stability of ecological systems". Annual Review of Ecology and Systematics. 4: 1–24. doi:10.1146/annurev.es.04.110173.000245. JSTOR 2096802.

45. Kerr, B.; Riley, M. A.; Feldman, M. W.; Bohannan, B. J. M. (2002). "Local dispersal promotes biodiversity in a real-life game of rock-paper-scissors". Nature. 418 (6894): 171–174. doi:10.1038/nature00823. PMID 12110887.

46. Law, R.; Morton, R. D. (1993). "Alternative permanent states of ecological communities". Ecology. 74 (5): 1347–1361. doi:10.2307/1940065. JSTOR

1940065.

47. Lewontin, R. C. (1969). "The meaning of stability". Brookhaven Symposia in Biology. 22: 13–23.
48. May, R. M. (1977). "Thresholds and breakpoints in ecosystems with a multiplicity of states". Nature. 267 (5628): 471–477. doi:10.1038/269471a0.
49. Mumby, P. J.; Hastings, A.; Edwards, H.J. (2007). "Thresholds and the resilience of Caribbean coral reefs". Nature. 450 (7166): 98–101. doi:10.1038/nature06252. PMID 17972885.
50. Paine, R. T. (1966). "Food web complexity and species diversity". The American Naturalist. 100 (910): 65–75. doi:10.1086/282400. JSTOR 2459379.
51. Peterson, G. D. (2002). "Forest dynamics in the Southeastern United States: managing multiple stable states". See Gunderson and Pritchard 2002, pp. 227–246.
52. Schröder, A.; Persson, L.; De Roos, A. M. (2005). "Direct experimental evidence for alternative stable states: a review". Oikos. 110 (1): 3–19. doi:10.1111/j.0030-1299.2005.13962.x.
53. Scheffer, M. (1997). Ecology of Shallow Water Lakes. London: Chapman and Hall. ISBN 0-412-74920-3.
54. Scheffer, M.; Carpenter, S. R. (2003). "Catastrophic regime shifts in ecosystems: linking theory to observation". Trends in Ecology and Evolution. 18 (12): 648–656. doi:10.1016/j.tree.2003.09.002.
55. Scheffer, M.; Carpenter, S.; Foley, J. A.; Folke, C.; Walker, B. (2001). "Catastrophic shifts in ecosystems". Nature. 413 (6856): 591–596. doi:10.1038/35098000. PMID 11595939.
56. Schmitz, O. J. (2004). "Perturbation and abrupt shift in trophic control of biodiversity and productivity". Ecology Letters. 7 (5): 403–409. doi:10.1111/j.1461-0248.2004.00592.x.
57. Sinclaire, A. R. E.; Krebs, C. J. (2002). "Complex numerical responses to top-down and bottom-up processes in vertebrate populations". Philosophical Transactions of the Royal Society B. 357 (1425): 1221–1231. doi:10.1098/rstb.2002.1123.
58. Sutherland, J. P. (1974). "Multiple stable points in natural communities". The American Naturalist. 108 (964): 859–873. doi:10.1086/282961. JSTOR 2459615.
59. Wilson, J. B.; Agnew, A. D. Q. (1992). "Positive-feedback switches in plant communities". Advances in Ecological Research. 23: 263–336. doi:10.1016/S0065-2504(08)60149-X.

Ecological Economics

Chapter Contents

Ecological economics is a trans-disciplinary field. It's not trying to be a subdiscipline of economics or a subdiscipline of ecology, but really it's a bridge across not only ecology and economics but also psychology, anthropology, archaeology, and history. That's what's necessary to get a more integrated picture of how humans have interacted with their environment in the past and how they might interact in the future. It's an attempt to look at humans embedded in their ecological life-support system, not separate from the environment. It also has some design elements, in the sense of how do we design a sustainable future.

INTRODUCTION

Ecological economics/eco-economics (also called Ecolonomy) refers to both a transdisciplinary and interdisciplinary field of academic research that aims to address the interdependence and coevolution of human economies and natural ecosystems over time and space. It is distinguished from environmental economics, which is the mainstream economic analysis of the environment, by its treatment of the economy as a property of the ecosystem, and its emphasis upon preserving natural capital. One survey of German economists found that ecological and environmental economics are different schools of economic thought, with ecological economists emphasizing strong sustainability and rejecting the proposition that natural capital can be substituted by human-made capital.

Ecological economics was founded as a modern movement in the works of and interactions between various European and American academics (see the section on history and development below). The related field of green economics is, in general, a more politically applied form of the subject.

According to ecological economist Malte Faber, ecological economics is defined by its focus on nature, justice, and time. Issues of intergenerational equity, irreversibility of environmental change, uncertainty of long-term outcomes, and sustainable development guide ecological economic analysis and valuation.Ecological economists have questioned fundamental mainstream economic approaches such as cost-benefit analysis, and the separability of economic values from scientific research, contending that economics is unavoidably normative rather than positive (i.e. descriptive). Positional analysis, which attempts to incorporate time and justice issues, is proposed as an alternative. Ecological economics shares many of its perspectives with feminist economics, including the focus on sustainability, nature, justice and care values.

History and Development

The first principles, deriving from the radiochemist FA Soddy, were laid out in his 1926 book "Wealth, Money (Virtual Wealth) and Debt" in 1926. Early modern interest in ecology and economics dates back to the 1940s in the work of K. William Kapp and Karl Polanyi and the 1960s in work by Kenneth Boulding and Herman Daly. However, the first organized meetings of modern ecological economists occurred in the 1980s. These began in 1982, at the instigation of Lois Banner, with a meeting held in Sweden (including Robert Costanza, Herman Daly, Charles Hall, Bruce Hannon, H.T. Odum, and David Pimentel). Most were ecosystem ecologists or mainstream environmental economists, with the exception of Daly. In 1987, Daly and Costanza edited an issue of *Ecological Modeling* to test the waters. A book entitled *Ecological Economics*, by Juan Martinez-Alier, was published later that year. 1989 saw the foundation of the International Society for Ecological Economics and publication of its journal, *Ecological Economics*, by Elsevier. Robert Costanza was the first president of the society and first editor of the journal, currently edited by Richard Howarth.

European conceptual founders include Nicholas Georgescu-Roegen (1971), K. William Kapp (1950) and Karl Polanyi (1944). Some key concepts of what is now ecological economics are evident in the writings of E.F. Schumacher, whose book *Small Is Beautiful – A Study of Economics as if People Mattered* (1973) was published just a few years before the first edition of Herman Daly›s comprehensive and persuasive *Steady-State Economics* (1977). Other figures include ecologists C.S. Holling, H.T. Odum and Robert Costanza, biologist Gretchen Daily and physicist Robert Ayres. CUNY geography professor David Harvey explicitly added ecological concerns to political economic literature. This parallel development in political economy has been continued by analysts such as sociologist John Bellamy Foster.

The antecedents can be traced back to the Romantics of the 19th century as well as some Enlightenment political economists of that era. Concerns over population were expressed by Thomas Malthus, while John Stuart Mill predicted the desirability of the *stationary state* of an economy. Mill thereby anticipated later insights of modern ecological economists, but without having had their experience of the social and ecological costs of the Post–World War II economic expansion. As Martinez-Alier explores in his book the debate on energy in economic systems can also be traced into the 19th century e.g. Nobel prize-winning chemist, Frederick Soddy (1877–1956). Soddy criticized the prevailing belief of the economy as a perpetual motion machine, capable of generating infinite wealth—a criticism echoed by his intellectual heirs in the now emergent field of ecological economics.

The Romanian economist Nicholas Georgescu-Roegen (1906–1994), who was among Daly's teachers at Vanderbilt University, provided ecological economics with a modern conceptual framework based on the material and energy flows of economic production and consumption. His *magnum opus*, *The Entropy Law and the Economic Process* (1971), has been highly influential.

Articles by Inge Ropke (2004, 2005) and Clive Spash (1999) cover the development and modern history of ecological economics and explain its differentiation from resource and environmental economics, as well as some of the controversy between American and European schools of thought. An article by Robert Costanza, David Stern, Lining He, and Chunbo Ma responded to a call by Mick Common to determine the foundational literature of ecological economics by using citation analysis to examine which books and articles have had the most influence on the development of the field. However, citations analysis has itself proven controversial and similar work has been criticized by Clive Spash for attempting to pre-determine what is regarded as influential in ecological economics through study design and data manipulation. In addition, the journal Ecological Economics has itself been criticized for swamping the field with mainstream economics.

NATURE AND ECOLOGY

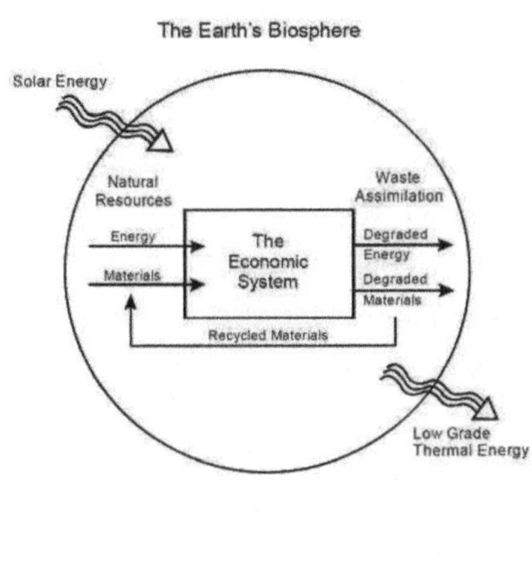

Natural resources flow through the economy and end up as waste and pollution

A simple circular flow of income diagram is replaced in ecological economics by a more complex flow diagram reflecting the input of solar energy, which sustains natural inputs and environmental services which are then used as units of production. Once consumed, natural inputs pass out of the economy as pollution and waste. The potential of an environment to provide services and materials is referred to as an environment›s source function, and this function is depleted as resources are consumed or pollution contaminates the resources. The sink function describes an environment›s ability to absorb and render harmless waste and pollution: when waste output exceeds the limit of the sink function, long-term damage occurs. Some persistent pollutants, such as some organic pollutants and nuclear waste are absorbed very slowly or not at all; ecological economists emphasize minimizing "cumulative pollutants". Pollutants affect human health and the health of the climate.

The economic value of natural capital and ecosystem services is accepted by mainstream environmental economics, but is emphasized as especially important in ecological economics. Ecological economists may begin by estimating how to maintain a stable environment before assessing the cost in dollar terms. Ecological economist Robert Costanza led an attempted valuation of the global ecosystem in 1997. Initially published in *Nature*, the article concluded on \$33 trillion with a range from \$16 trillion to \$54 trillion (in 1997, total global GDP was \$27 trillion). Half of the value went to nutrient cycling. The open oceans, continental shelves, and estuaries had the highest total value, and the highest per-hectare values went to estuaries, swamps/floodplains, and seagrass/algae beds. The work was criticized by articles in *Ecological Economics* Volume 25, Issue 1, but the critics acknowledged the positive potential for economic valuation of the global ecosystem.

The Earth's carrying capacity is a central issue in ecological economics. Early economists such as Thomas Malthus pointed out the finite carrying capacity of the earth, which was also central to the MIT study *Limits to Growth*. Diminishing returns suggest that productivity increases will slow if major technological progress is not made. Food production may become a problem, as erosion, an impending water crisis, and soil salinity (from irrigation) reduce the productivity of agriculture. Ecological economists argue that industrial agriculture, which exacerbates these problems, is not sustainable agriculture, and are generally inclined favorably to organic farming, which also reduces the output of carbon.

Global wild fisheries are believed to have peaked and begun a decline, with valuable habitat such as estuaries in critical condition. The aquaculture or farming of piscivorous fish, like salmon, does not help solve the problem because they need to be fed products from other fish. Studies have shown that salmon farming has major negative impacts on wild salmon, as well as the forage fish that need to be caught to feed them.

Since animals are higher on the trophic level, they are less efficient sources of food energy. Reduced consumption of meat would reduce the demand for food, but as nations develop, they tend to adopt high-meat diets similar to that of the United States. Genetically modified food (GMF) a conventional solution to the problem, presents numerous problems – Bt corn produces its own Bacillus thuringiensis toxin/protein, but the pest resistance is believed to be only a matter of time. The overall effect of GMF on yields is contentious, with the USDA and FAO acknowledging that GMFs do not necessarily have higher yields and may even have reduced yields.

Global warming is now widely acknowledged as a major issue, with all national scientific academies expressing agreement on the importance of the issue. As the population growth intensifies and energy demand increases, the world faces an energy crisis. Some economists and scientists forecast a global ecological crisis if energy use is not contained – the Stern report is an example. The disagreement has sparked a vigorous debate on issue of discounting and intergenerational equity.

GLOBAL GEOCHEMICAL CYCLES CRITICAL FOR LIFE

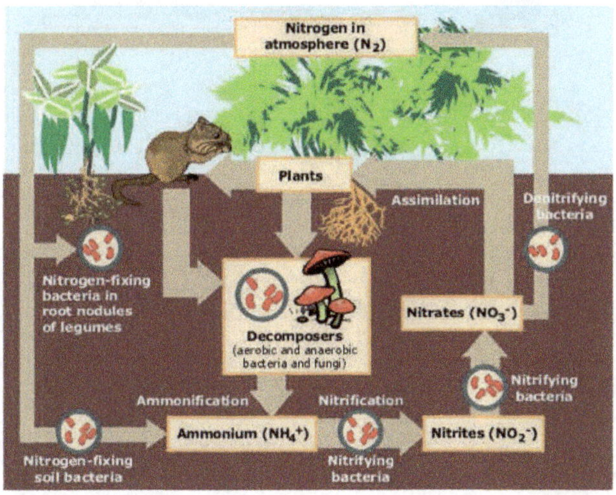

Nitrogen cycle

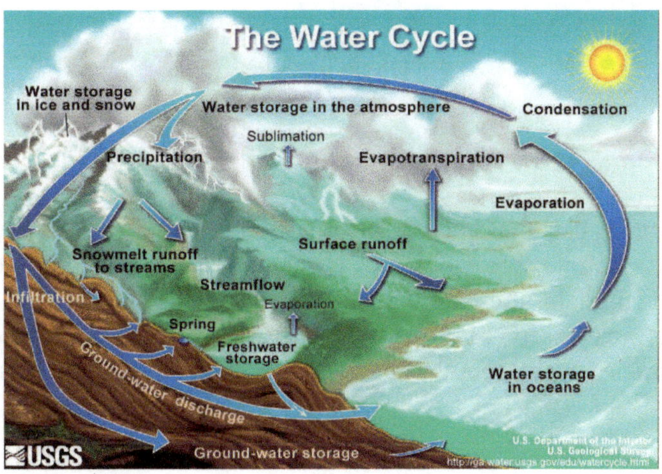

Water cycle

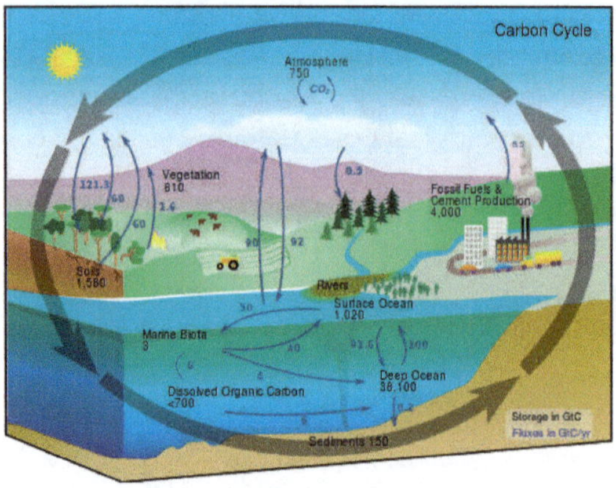

Carbon cycle

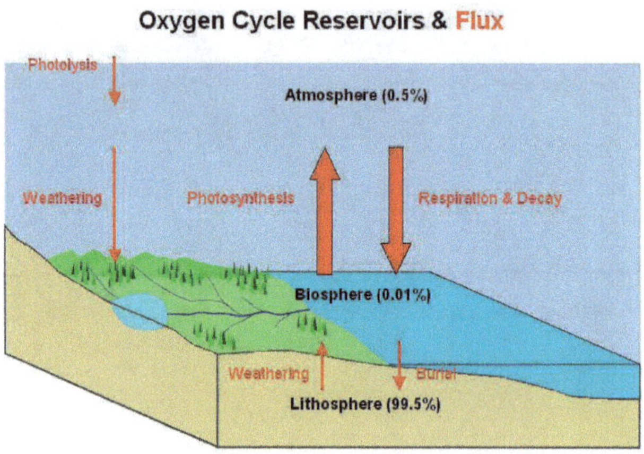

Oxygen cycle

ETHICS

Mainstream economics has attempted to become a value-free 'hard science›, but ecological economists argue that value-free economics is generally not realistic. Ecological economics is more willing to entertain alternative conceptions of utility, efficiency, and cost-benefits such as positional analysis or multi-criteria analysis. Ecological economics is typically viewed as economics for sustainable development, and may have goals similar to green politics.

SCHOOLS OF THOUGHT

Various competing schools of thought exist in the field. Some are close to resource and environmental economics while others are far more heterodox in outlook. An example of the latter is the *European Society for Ecological Economics*. An example of the former is the Swedish *Beijer International Institute of Ecological Economics.* Clive Spash has argued for the classification of the ecological economics movement, and more generally work by different economic schools on the environment, into three main categories. These are the mainstream new resource economists, the new environmental pragmatists, and the more radical social ecological economists. International survey work comparing the relevance of the categories for mainstream and heterodox economists shows some clear divisions between environmental and ecological economists.

DIFFERENTIATION FROM MAINSTREAM SCHOOLS

Some ecological economists prioritise adding natural capital to the typical capital asset analysis of land, labor, and financial capital. These ecological economists then use tools from mathematical economics as in mainstream economics, but may apply them more closely to the natural world. Whereas mainstream economists tend to be technological optimists, ecological economists are inclined to be technological sceptics. They reason that the natural world has a limited carrying capacity and that its resources may run out. Since destruction of important environmental resources could be practically irreversible and catastrophic, ecological economists are inclined to justify cautionary measures based on the precautionary principle.

The most cogent example of how the different theories treat similar assets is tropical rainforest ecosystems, most obviously the Yasuni region of Ecuador. While this area has substantial deposits of bitumen it is also one of the most diverse ecosystems on Earth and some estimates establish it has over 200 undiscovered medical substances in its genomes - most of which would be destroyed by logging the forest or mining the bitumen. Effectively, the instructional capital of the genomes is undervalued by analyses which view the rainforest primarily as a source of wood, oil/tar and perhaps food. Increasingly the carbon credit for leaving the extremely carbon-intensive ("dirty") bitumen in the ground is also valued - the government of Ecuador set a price of US$350M for an oil lease with the intent of selling it to someone committed to never exercising it at all and instead preserving the rainforest.

While this natural capital and ecosystems services approach has proven popular amongst many it has also been contested as failing to address the underlying problems with mainstream economics, growth, market capitalism and monetary valuation of the environment. Critiques concern the need to create a more meaningful relationship with Nature and the non-human world than evident in the instrumentalism of shallow ecology and the environmental economists commodification of everything external to the market system.

TOPICS

Among the topics addressed by ecological economics are methodology, allocation of resources, weak versus strong sustainability, energy economics, energy accounting and balance, environmental services, cost shifting, and modeling.

METHODOLOGY

A primary objective of ecological economics (EE) is to ground economic thinking and practice in physical reality, especially in the laws of physics (particularly the laws of thermodynamics) and in knowledge of biological systems. It accepts as a goal the improvement of human well-being through development, and seeks to ensure achievement of this through planning for the sustainable development of ecosystems and societies. Of course the terms development and sustainable development are far from lacking controversy. Richard B. Norgaard argues traditional economics has hi-jacked the development terminology in his book *Development Betrayed*.

Well-being in ecological economics is also differentiated from welfare as found in mainstream economics and the 'new welfare economics' from the 1930s which informs resource and environmental economics. This entails a limited preference utilitarian conception of value i.e., Nature is valuable to our economies, that is because people will pay for its services such as clean air, clean water, encounters with wilderness, etc.

Ecological economics is distinguishable from neoclassical economics primarily by its assertion that the economy is embedded within an environmental system. Ecology deals with the energy and matter transactions of life and the Earth, and the human economy is by definition contained within this system. Ecological economists argue that neoclassical economics has ignored the environment, at best considering it to be a subset of the human economy.

The neoclassical view ignores much of what the natural sciences have taught us about the contributions of nature to the creation of wealth e.g., the planetary endowment of scarce matter and energy, along with the complex and biologically diverse ecosystems that provide goods and ecosystem services directly to human communities: micro- and macro-climate regulation, water recycling, water purification, storm water regulation, waste absorption, food and medicine production, pollination, protection from solar and cosmic radiation, the view of a starry night sky, etc.

There has then been a move to regard such things as natural capital and ecosystems functions as goods and services.However, this is far from uncontroversial within ecology or ecological economics due to the potential for narrowing down values to those found in mainstream economics and the danger of merely regarding Nature as a commodity. This has been referred to as ecologists ‹selling out on Nature›. There is then a concern that ecological economics has failed to learn from the extensive literature in environmental ethics about how to structure a plural value system.

Allocation of Resources

Resource and neoclassical economics focus primarily on the efficient allocation of resources, and less on two other fundamental economic problems which are central to ecological economics: distribution (equity) and the scale of the economy relative to the ecosystems upon which it is reliant. Ecological Economics also makes a clear distinction between growth (quantitative increase in economic output) and development (qualitative improvement of the quality of life) while arguing that neoclassical economics confuses the two. Ecological economists point out that, beyond modest levels, increased per-capita consumption (the typical economic measure of "standard of living") does not necessarily lead to improvement in human well-being, while this same consumption can have harmful effects on the environment and broader societal well-being.

Weak Versus Strong Sustainability

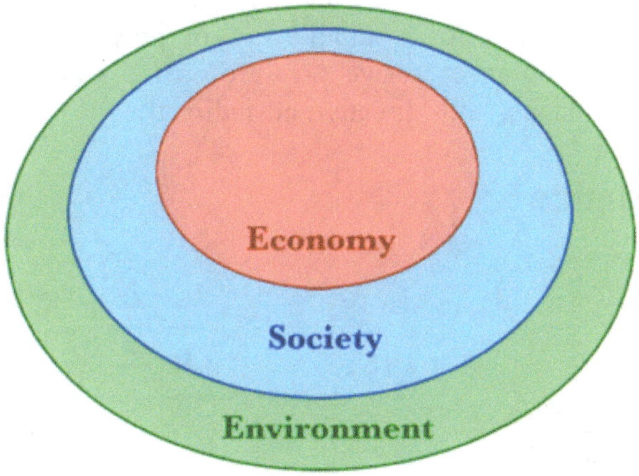

The three nested systems of sustainability - the economy wholly contained by society, wholly contained by the biophysical environment. Clickable.

Ecological economics challenges the conventional approach towards natural resources, claiming that it undervalues natural capital by considering it as interchangeable with human-made capital—labor and technology.

The impending depletion of natural resources and increase of climate-changing greenhouse gasses should motivate us to examine how political, economic and social policies can benefit from alternative energy. Interestingly enough, shifting our dependence on fossil fuels with specific interest within just one of the above-mentioned factors easily benefits at least one other. For instance, photo voltaic (or solar) panels have a 15% efficiency when absorbing the sun›s energy, but its construction demand has increased 120% within both commercial and residential properties. Additionally, this construction has led to a roughly 30% increase in work demands (Chen).

The potential for the substitution of man-made capital for natural capital is an important debate in ecological economics and the economics of sustainability. There is a continuum of views among economists between the strongly neoclassical positions of Robert Solow and Martin Weitzman, at one extreme and the 'entropy pessimists', notably Nicholas Georgescu-Roegen and Herman Daly, at the other.

Neoclassical economists tend to maintain that man-made capital can, in principle, replace all types of natural capital. This is known as the weak sustainability view, essentially that every technology can be improved upon or replaced by innovation, and that there is a substitute for any and all scarce materials.

At the other extreme, the strong sustainability view argues that the stock of natural resources and ecological functions are irreplaceable. From the premises of strong sustainability, it follows that economic policy has a fiduciary responsibility to the greater ecological world, and that sustainable development must therefore take a different approach to valuing natural resources and ecological functions.

Recently, Stanislav Shmelev developed a new methodology for the assessment of progress at the macro scale based on multi-criteria methods, which allows consideration of different perspectives, including strong and weak sustainability or conservationists vs industrialists and aims to search for a ‹middle way› by providing a strong neo-Keynsian economic push without putting excessive pressure on the natural resources, including water or producing emissions, both directly and indirectly.

Energy Economics

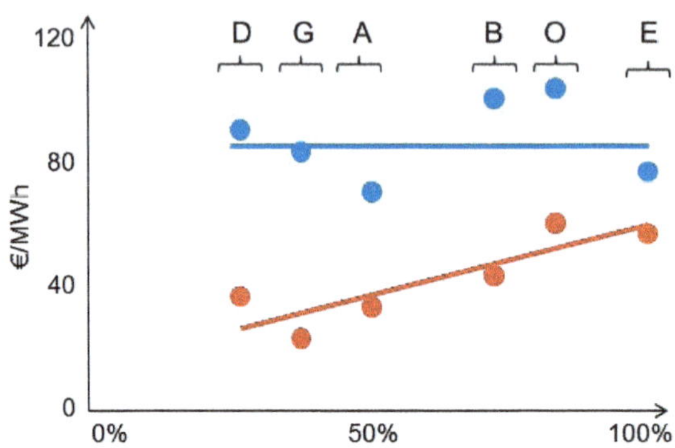

Exergy analysis can be performed to find connections between economic value and the physical world. Here the costs of heating (vertical axis) are compared with the exergy content of different energy carriers (horizontal axis). Red dots and trend line indicate energy prices for consumers, blue dots and trend line indicate total price for consumers including capital expenditure for the heating system. Energy carriers included are district heating (D), ground-source heat pump (G), exhaust air heat pump (A), bioenergy meaning firewood (B), heating oil (O) and direct electric heating (E).

A key concept of energy economics is net energy gain, which recognizes that all energy requires energy to produce. To be useful the energy return on energy invested (*EROEI*) has to be greater than one. The net energy gain from production coal, oil and gas has declined over time as the easiest to produce sources have been most heavily depleted.

Ecological economics generally rejects the view of energy economics that growth in the energy supply is related directly to well being, focusing instead on biodiversity and creativity - or natural capital and individual capital, in the terminology sometimes adopted to describe these economically. In practice, ecological economics focuses primarily on the key issues of uneconomic growth and quality of life. Ecological economists are inclined to acknowledge that much of what is important in human well-being is not analyzable from a strictly economic standpoint and suggests an interdisciplinary approach combining social and natural sciences as a means to address this.

Thermoeconomics is based on the proposition that the role of energy in biological evolution should be defined and understood through the second law of thermodynamics, but also in terms of such economic criteria as productivity, efficiency, and especially the costs and benefits (or profitability) of the various mechanisms for capturing and utilizing available energy to build biomass and do work. As a result, thermoeconomics is often discussed in the field of ecological economics, which itself is related to the fields of sustainability and sustainable development.

Exergy analysis is performed in the field of industrial ecology to use energy more efficiently. The term *exergy*, was coined by Zoran Rant in 1956, but the concept was developed by J. Willard Gibbs. In recent decades, utilization of exergy has spread outside of physics and engineering to the fields of industrial ecology, ecological economics, systems ecology, and energetics.

Energy Accounting and Balance

An energy balance can be used to track energy through a system, and is a very useful tool for determining resource use and environmental impacts, using the First and Second laws of thermodynamics, to determine how much energy is needed at each point in a system, and in what form that energy is a cost in various environmental issues. The energy accounting system keeps track of energy in, energy out, and non-useful energy versus work done, and transformations within the system.

Scientists have written and speculated on different aspects of energy accounting.

Environmental Services

A study was carried out by Costanza and colleagues to determine the 'price' of the services provided by the environment. This was determined by averaging values obtained from a range of studies conducted in very specific context and then transferring these without regard to that context. Dollar figures were averaged to a per hectare number for different types of ecosystem e.g. wetlands, oceans. A total was then produced which came out at 33 trillion US dollars (1997 values), more than twice the total GDP of the world at the time of the study. This study was criticized by pre-ecological and even some environmental economists - for being inconsistent with assumptions of financial capital valuation - and ecological economists - for being inconsistent with an ecological economics focus on biological and physical indicators.

The whole idea of treating ecosystems as goods and services to be valued in monetary terms remains controversial. A common objection is that life is precious or priceless, but this demonstrably degrades to it being worthless under the assumptions of any branch of economics. Reducing human bodies to financial values is a necessary part of every branch of economics and not always in the direct terms of insurance or wages. Economics, in principle, assumes that conflict is reduced by agreeing on voluntary contractual relations and prices instead of simply fighting or coercing or tricking others into providing goods or services. In doing so, a provider agrees to surrender time and take bodily risks and other (reputation, financial) risks. Ecosystems are no different from other bodies economically except insofar as they are far less replaceable than typical labour or commodities.

Despite these issues, many ecologists and conservation biologists are pursuing ecosystem valuation. Biodiversity measures in particular appear to be the most promising way to reconcile financial and ecological values, and there are many active efforts in this regard. The growing field of biodiversity finance began to emerge in 2008 in response to many specific proposals such as the Ecuadoran Yasuni proposal or similar ones in the Congo. US news outlets treated the stories as a threat to drill a park reflecting a previously dominant view that NGOs and governments had the primary responsibility to protect ecosystems. However Peter Barnes and other commentators have recently argued that a guardianship/trustee/commons model is far more effective and takes the decisions out of the political realm.

Commodification of other ecological relations as in carbon credit and direct payments to farmers to preserve ecosystem services are likewise examples that enable private parties to play more direct roles protecting biodiversity. The United NationsFood and Agriculture Organization achieved near-universal agreement in 2008 that such payments directly valuing ecosystem preservation and encouraging permaculture were the only practical way out of a food crisis. The holdouts were all English-speaking countries that export GMOs and promote "free trade agreements that facilitate their own control of the world transport network: The US, UK, Canada and Australia.

Not 'externalities', But Cost shifTing

Ecological economics is founded upon the view that the neoclassical economics (NCE) assumption that environmental and community costs and benefits are mutually canceling *"externalities"* is not warranted. Joan Martinez Alier, for instance shows that the bulk of consumers are automatically excluded from having an impact upon the prices of commodities, as these consumers are future generations who have not been born yet. The assumptions behind future discounting, which assume that future goods will be cheaper than present goods, has been criticized by Fred Pearce and by the recent Stern Report (although the Stern report itself does employ discounting and has been criticized for this and other reasons by ecological economists such as Clive Spash).

Concerning these externalities, some like the eco-businessman Paul Hawken argue an orthodox economic line that the only reason why goods produced unsustainably are usually cheaper than goods produced sustainably is due to a hidden subsidy, paid by the non-monetized human environment, community or future generations. These arguments are developed further by Hawken, Amory and Hunter Lovins to promote their vision of an environmental capitalist utopia in *Natural Capitalism: Creating the Next Industrial Revolution*.

In contrast, ecological economists, like Joan Martinez-Alier, appeal to a different line of reasoning. Rather than assuming some (new) form of capitalism is the best way forward, an older ecological economic critique questions the very idea of internalizing externalities as providing some corrective to the current system. The work by Karl William Kapp explains why the concept of "externality" is a misnomer. In fact the modern business enterprise operates on the basis of shifting costs onto others as normal practice to make profits. Charles Einsentein has argued that this method of privatising profits while socialising the costs through externalities, passing the costs to the community, to the natural environment or to future generations is inherently destructive As social ecological economist Clive Spash has noted, externality theory fallaciously assumes environmental and social problems are minor aberrations in an otherwise perfectly functioning efficient economic system. Internalizing the odd externality does nothing to address the structural systemic problem and fails to recognize the all pervasive nature of these supposed ‹externalities›.

Ecological-Economic Modeling

Mathematical modeling is a powerful tool that is used in ecological economic analysis. Various approaches and techniques include: evolutionary, input-output, neo-Austrian modeling, entropy and thermodynamic models, multi-criteria, and agent-based modeling, the environmental Kuznets curve, and Stock-Flow consistent model frameworks. System dynamics and GIS are techniques applied, among other, to spatial dynamic landscape simulation modeling. The Matrix accounting methods of Christian Felber provide a more sophisticated method for identifying the common good

CRITICISM

Assigning monetary value to natural resources such as biodiversity, and the emergent ecosystem services is often viewed as a key process in influencing economic practices, policy, and decision-making. While this idea is becoming more and more accepted among ecologists and conservationist, some argue that it is inherently false.

McCauley argues that ecological economics and the resulting ecosystem service based conservation can be harmful. He describes four main problems with this approach:

Firstly, it seems to be assumed that all ecosystem services are financially beneficial. This is undermined by a basic characteristic of ecosystems: they do not act specifically in favour of any single species. While certain services might be very useful to us, such as coastal protection from hurricanes by mangroves for example, others might cause financial or personal harm, such as wolves hunting cattle. The complexity of Eco-systems makes it challenging to weigh up the value of a given species. Wolves play a critical role in regulating prey populations, the absence of such an apex predator in the Scottish Highlands have caused the over population of deer, preventing afforestation, which increases the risk of flooding and damage to property.

Secondly, allocating monetary value to nature would make its conservation reliant on markets that fluctuate. This can lead to devaluation of services that were previously considered financially beneficial. Such is the case of the bees in a forest near former coffee plantations in Finca Santa Fe, Costa Rica. The pollination services were valued to over US$60,000 a year, but soon after the study, coffee prices dropped and the fields were replanted with pineapple. Pineapple does not require bees to be pollinated, so the value of their service dropped to zero.

Thirdly, conservation programmes for the sake of financial benefit underestimate human ingenuity to invent and replace ecosystem services by artificial means. McCauley argues that such proposals are deemed to have a short lifespan as the history of technology is about how Humanity developed artificial alternatives to nature's services and with time passing the cost of such services tend to decrease. This would also lead to the devaluation of ecosystem services.

Lastly, it should not be assumed that conserving ecosystems is always financially beneficial as opposed to alteration. In the case of the introduction of the Nile perch to Lake Victoria, the ecological consequence was decimation of native fauna. However, this same event is praised by the local communities as they gain significant financial benefits from trading the fish.

McCauley argues that, for these reasons, trying to convince decision-makers to conserve nature for monetary reasons is not the path to be followed, and instead appealing to morality is the ultimate way to campaign for the protection of nature.

ECOLOGICAL FOOTPRINT

The natural resources of the Earth are finite.

The **ecological footprint** measure how much nature it takes to support people. It is an ecological accounting system. It contrasts how much biologically productive area people use for their consumption to how much biologically productive area is available (biocapacity).

More specifically, the ecological footprint is the biologically productive area needed to provide for everything people use: fruits and vegetables, fish, wood, fibers, absorption of carbon dioxide from fossil fuel use, and space for buildings and roads. Biocapacity is the productive area that can regenerate what people demand from nature. Footprint and biocapacity can be compared at the individual, regional, national or global scale. Both footprint and biocapacity change every year with number of people, per person consumption, efficiency of production, and productivity of ecosystems.

At a global scale, footprint assessments show how big humanity›s demand is compared to what planet Earth can renew. Global Footprint Network calculates the global ecological footprint from UN and other data for over 200 nations. They estimate that as of 2012 humanity has been using natural capital 1.6 times as fast as nature can renew it.

Ecological footprint analysis is widely applied around the Earth as an indicator of environmental sustainability. It can be used to measure and manage the use of resources throughout the economy and explore the sustainability of individual lifestyles, goods and services, organizations, industry sectors, neighborhoods, cities, regions and nations. Since 2006, a first set of ecological footprint standards exist that detail both communication and calculation procedures.

Footprint Measurements

In 2007, the Global Footprint Network estimated the global ecological footprint as 1.6 planet Earths; that is, they judged that ecological services were being used 1.6 times as quickly as they were being renewed.

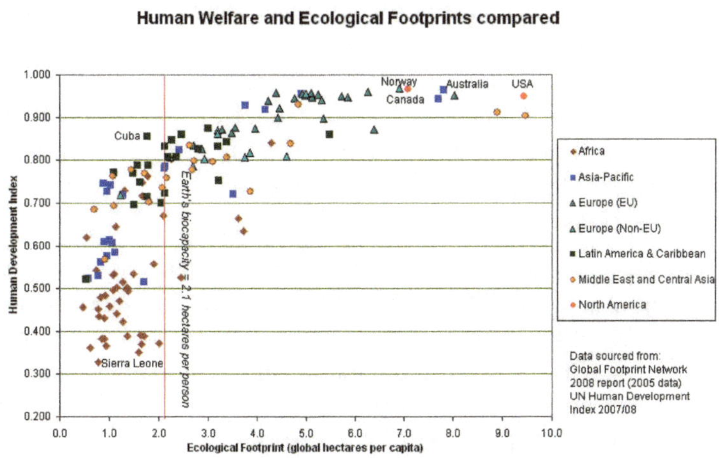

Ecological footprint for different nations compared to their Human Development Index.

Ecological footprints can be calculated at any scale: for an activity, a person, a community, a city, a region, a nation or humanity as a whole. Cities, due to population concentration, have large ecological footprints and have become ground zero for footprint reduction.

Global Footprints: Currently there is no fixed way to measure global footprints, and any attempts to describe the capacity of an ecosystem in a single number is a massive simplification of thousands of key renewable resources, which are not used or replenished at the same rate. However, there has been some convergence of metrics and standards since 2006.

City Ecological Footprints: are being measured. There are two types of measurements in use. The first measures ecosystem displacement which is defined as City Area minus remaining green spaces. This is an area measurement that does not include human or other biological activity. The Second attempts to quantify surviving ecosystem health.

Specifically, it attempts to quantify both area and biological health of ecosystems surviving inside city areas such as nature reserves, parks, other green spaces. City footprints are being calculated and ranked with city ecological indexes.

Overview

The first academic publication about ecological footprints was by William Rees in 1992. The ecological footprint concept and calculation method was developed as the PhD dissertation of Mathis Wackernagel, under Rees› supervision at the University of British Columbia in Vancouver, Canada, from 1990–1994. Originally, Wackernagel and Rees called the concept appropriated carrying capacity. To make the idea more accessible, Rees came up with the term ecological footprint, inspired by a computer technician who praised his new computer›s small footprint on the desk. In early 1996, Wackernagel and Rees published the book *Our Ecological Footprint: Reducing Human Impact on the Earth* with illustrations by Phil Testemale.

Footprint values at the end of a survey are categorized for Carbon, Food, Housing, and Goods and Services as well as the total footprint number of Earths needed to sustain the world›s population at that level of consumption. This approach can also be applied to an activity such as the manufacturing of a product or driving of a car. This resource accounting is similar to life-cycle analysis wherein the consumption of energy, biomass (food, fiber), building material, water and other resources are converted into a normalized measure of land area called global hectares (gha).

Per capita ecological footprint (EF), or ecological footprint analysis (EFA), is a means of comparing consumption and lifestyles, and checking this against nature›s ability to provide for this consumption. The tool can inform policy by examining to what extent a nation uses more (or less) than is available within its territory, or to what extent the nation›s lifestyle would be replicable worldwide. The footprint can also be a useful tool to educate people about carrying capacity and overconsumption, with the aim of altering personal behavior. Ecological footprints may be used to argue that many current lifestyles are not sustainable. Such a global comparison also clearly shows the inequalities of resource use on this planet at the beginning of the twenty-first century.

In 2007, the average biologically productive area per person worldwide was approximately 1.8 global hectares (gha) per capita. The U.S. footprint per capita was 9.0 gha, and that of Switzerland was 5.6 gha, while China›s was 1.8 gha. The WWF claims that the human footprint has exceeded the biocapacity (the available supply of natural resources) of the planet by 20%. Wackernagel and Rees originally estimated that the available biological capacity for the 6 billion people on Earth at that time was about 1.3 hectares per person, which is smaller than the 1.8 global hectares published for 2006, because the initial studies neither used global hectares nor included bioproductive marine areas.

A number of NGOs offer ecological footprint calculators (*see* Footprint Calculator, below).

Methodology

The ecological footprint accounting method at the national level is described in the l Footprint Atlas 2010 or in greater detail in the Calculation Methodology for the National Footprint Accounts. The National Accounts Review Committee has also published a research agenda on how the method will be improved.

In 2003, Jason Venetoulis, Carl Mas, Christopher Gaudet, Dahlia Chazan, and John Talberth developed Footprint 2., which offers a series of theoretical and methodological improvements to the standard footprint approach. The four primary improvements were that they included the entire surface of the Earth in biocapacity estimates, allocated space for other (i.e., non-human) species, updated the basis of equivalence factors from agricultural land to net primary productivity (NPP), and refined the carbon component of the footprint based on the latest global carbon models.

Studies in the United Kingdom

The UK's average ecological footprint is 5.45 global hectares per capita (gha) with variations between regions ranging from 4.80 gha (Wales) to 5.56 gha (East England).

Two recent studies have examined relatively low-impact small communities. BedZED, a 96-home mixed-income housing development in South London, was designed by Bill Dunster Architects and sustainability consultants BioRegional for the Peabody Trust. Despite being populated by relatively mainstream home-buyers, BedZED was found to have a footprint of 3.20 gha due to on-site renewable energy production, energy-efficient architecture, and an extensive green lifestyles program that included on-site London›s first carsharing club. The report did not measure the added footprint of the 15,000 visitors who have toured BedZED since its completion in 2002. Findhorn Ecovillage, a rural intentional community in Moray, Scotland, had a total footprint of 2.56 gha, including both the many guests and visitors who travel to the community to undertake residential courses there and the nearby campus of Cluny Hill College. However, the residents alone have a footprint of 2.71 gha, a little over half the UK national average and one of the lowest ecological footprints of any community measured so far in the industrialized world. Keveral Farm, an organic farming community in Cornwall, was found to have a footprint of 2.4 gha, though with substantial differences in footprints among community members.

Critiques

Early criticism was published by van den Bergh and Verbruggen in 1999, which was updated in 2014. Another criticism was published in 2008. A more complete review commissioned by the Directorate-General for the Environment (European Commission) was published in June 2008.

A recent critique of the concept is due to Blomqvist et al., 2013a, with a reply from Rees and Wackernagel, 2013, and a rejoinder by Blomqvist et al., 2013b.

An additional strand of critique is due to Giampietro and Saltelli (2014a), with a reply from Goldfinger et al., 2014, a rejoinder by Giampietro and Saltelli (2014a), and additional comments from van den Bergh and Grazi (2015).

A number of countries have engaged in research collaborations to test the validity of the method. This includes Switzerland, Germany, United Arab Emirates, and Belgium.

Grazi et al. (2007) have performed a systematic comparison of the ecological footprint method with spatial welfare analysis that includes environmental externalities, agglomeration effects and trade advantages. They find that the two methods can lead to very distinct, and even opposite, rankings of different spatial patterns of economic activity. However this should not be surprising, since the two methods address different research questions.

Calculating the ecological footprint for densely populated areas, such as a city or small country with a comparatively large population — e.g. New York and Singapore respectively — may lead to the perception of these populations as parasitic. This is because these communities have little intrinsic biocapacity, and instead must rely upon large *hinterlands*. Critics argue that this is a dubious characterization since mechanized rural farmers in developed nations may easily consume more resources than urban inhabitants, due to transportation requirements and the unavailability of economies of scale. Furthermore, such moral conclusions seem to be an argument for autarky. Some even take this train of thought a step further, claiming that the Footprint denies the benefits of trade. Therefore, the critics argue that the Footprint can only be applied globally.

The method seems to reward the replacement of original ecosystems with high-productivity agricultural monocultures by assigning a higher biocapacity to such regions. For example, replacing ancient woodlands or tropical forests with monoculture forests or plantations may improve the ecological footprint. Similarly, if organic farming yields were lower than those of conventional methods, this could result in the former being "penalized" with a larger ecological footprint. Of course, this insight, while valid, stems from the idea of using the footprint as one's only metric. If the use of ecological footprints are complemented with other indicators, such as one for biodiversity, the problem could maybe be solved. Indeed, WWF›s Living Planet Report complements the biennial Footprint calculations with the Living Planet Index of biodiversity. Manfred Lenzen and Shauna Murray have created a modified Ecological Footprint that takes biodiversity into account for use in Australia.

Although the ecological footprint model prior to 2008 treated nuclear power in the same manner as coal power, the actual real world effects of the two are radically different. A life cycle analysis centered on the Swedish Forsmark Nuclear Power Plant estimated carbon dioxide emissions at 3.10 g/kWh and 5.05 g/kWh in 2002 for the Torness Nuclear Power Station. This compares to 11 g/kWh for hydroelectric power, 950 g/kWh for installed coal, 900 g/kWh for oil and 600 g/kWh for natural gas generation in the United States in 1999. Figures released by Mark Hertsgaard, however, show that because of the delays in building nuclear plants and the costs involved, investments in energy efficiency and renewable energies have seven times the return on investment of investments in nuclear energy.

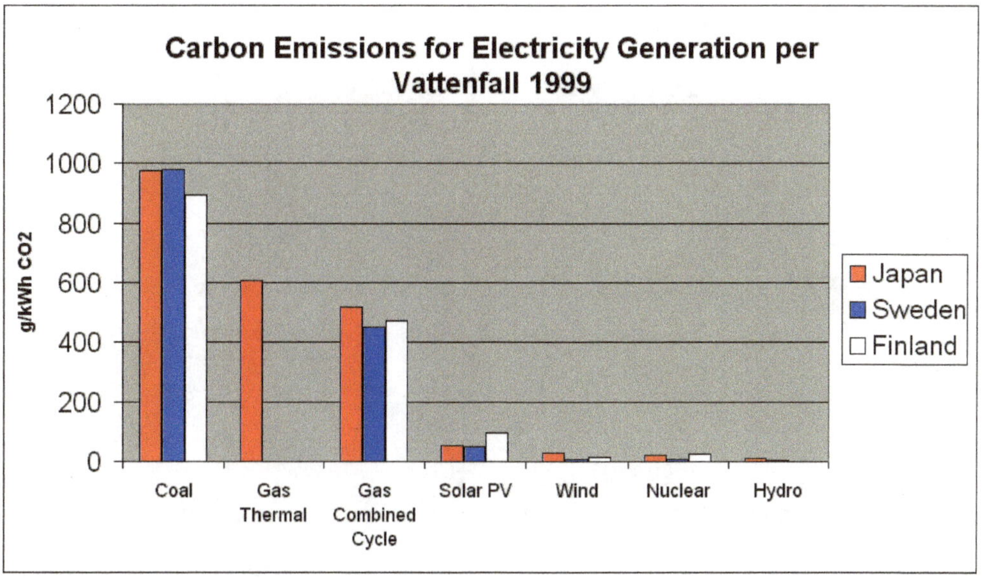

The Swedish utility Vattenfall did a study of full life-cycle greenhouse-gas emissions of energy sources the utility uses to produce electricity, namely: Nuclear, Hydro, Coal, Gas, Solar Cell, Peat and Wind. The net result of the study was that nuclear power produced 3.3 grams of carbon dioxide per KW-Hr of produced power. This compares to 400 for natural gas and 700 for coal (according to this study). The study also concluded that nuclear power produced the smallest amount of CO_2 of any of their electricity sources.

Claims exist that the problems of nuclear waste do not come anywhere close to approaching the problems of fossil fuel waste. A 2004 article from the BBC states: The World Health Organization (WHO) says 3 million people are killed worldwide by outdoor air pollution annually from vehicles and industrial emissions, and 1.6 million indoors through using solid fuel." In the U.S. alone, fossil fuel waste kills 20,000 people each year. A coal power plant releases 100 times as much radiation as a nuclear power plant of the same wattage. It is estimated that during 1982, US coal burning released 155 times as much radioactivity into the atmosphere as the Three Mile Island incident. In addition, fossil fuel waste causes global warming, which leads to increased deaths from hurricanes, flooding, and other weather events. The World Nuclear Association provides a comparison of deaths due to accidents among different forms of energy production. In their comparison, deaths per TW-yr of electricity produced (in UK and USA) from 1970 to 1992 are quoted as 885 for hydropower, 342 for coal, 85 for natural gas, and 8 for nuclear.

The Western Australian government State of the Environment Report included an Ecological Footprint measure for the average Western Australian seven times the average footprint per person on the planet in 2007, a total of about 15 hectares.

Footprint by Country

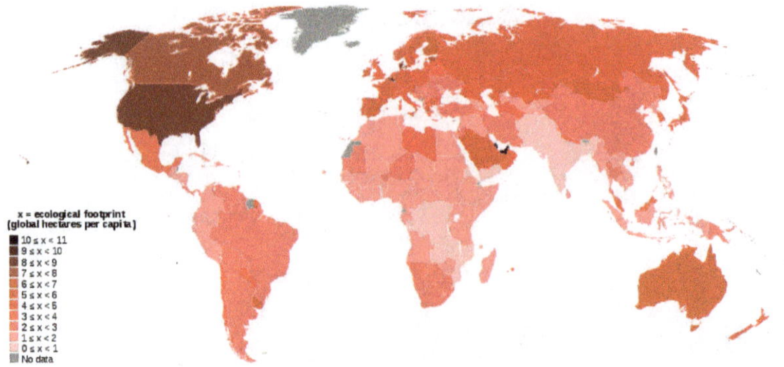

World map of countries by ecological footprint (data of 2007).

The world-average ecological footprint in 2012 was 1.8 global hectares per person. The average per country ranges from over 10 to under 1 hectares per person. There is also a high variation within countries, based on individual lifestyle and economic situation.

The GHG footprint differs from the ecological footprint in that the former is expressed in units of GHG warming potential (GGWP) and is generated by products or services, whereas the latter is expressed in units of land area and is generated by whole societies.

Implications

the average world citizen has an eco-footprint of about 2.7 global average hectares while there are only 2.1 global hectare of bioproductive land and water per capita on earth. This means that humanity has already overshot global biocapacity by 30% and now lives unsustainabily by depleting stocks of natural capital

REFERENCES

1. *Anastasios Xepapadeas (2008). Ecological economics. The New Palgrave Dictionary of Economics 2nd Edition. Palgrave MacMillan.*
2. Jeroen C.J.M. van den Bergh (2001). Ecological Economics: Themes, Approaches, and Differences with Environmental Economics, *Regional Environmental Change*, 2(1), pp. 13-23 (press +).
3. Illge L, Schwarze R. (2006). A Matter of Opinion: How Ecological and Neoclassical Environmental Economists Think about Sustainability and Economics . German Institute for Economic Research.
4. Paehlke R. (1995). *Conservation and Environmentalism: An Encyclopedia*, p. 315. Taylor & Francis.
5. Scott Cato, M. (2009). *Green Economics*. Earthscan, London. ISBN 978-1-84407-571-3.

6. Malte Faber. (2008). How to be an ecological economist. *Ecological Economics* **66**(1):1-7. Preprint.

7. Peter Victor. (2008). Book Review: Frontiers in Ecological Economic Theory and Application. *Ecological Economics***66**(2-3).

8. Mattson L. (1975). Book Review: *Positional Analysis for Decision-Making and Planning* by Peter Soderbaum. *The Swedish Journal of Economics*.

9. Soderbaum, P. 2008. *Understanding Sustainability Economics*. Earthscan, London. ISBN 978-1-84407-627-7. pp.109-110, 113-117.

10. *Aslaksen, Iulie; Bragstad, Torunn; Ås, Berit (2014). "Feminist Economics as Vision for a Sustainable Future". In Bjørnholt, Margunn; McKay, Ailsa. Counting on Marilyn Waring: New Advances in Feminist Economics. Demeter Press/Brunswick Books. pp. 21–36. ISBN 9781927335277.*

11. Røpke, I. (2004) The early history of modern ecological economics. Ecological Economics 50(3-4): 293-314.

12. Costanza R. (2003). Early History of Ecological Economics and ISEE. Internet Encyclopaedia of Ecological Economics.

13. Kapp, K. W. (1950) *The Social Costs of Private Enterprise*. New York: Shocken.

14. Polanyi, K. (1944) *The Great Transformation*. New York/Toronto: Rinehart & Company Inc.

15. Schumacher, E.F. 1973. *Small Is Beautiful: A Study of Economics as if People Mattered*. London: Blond and Briggs.

16. Daly, H. 1991. *Steady-State Economics* (2nd ed.). Washington, D.C.: Island Press.

17. Zencey, Eric. (2009, April 12). Op-ed. *New York Times*, p. WK9. Accessed: December 23, 2012.

18. Georgescu-Roegen, N. 1971. *The Entropy Law and the Economic Process*. Cambridge, Mass.: Harvard University Press.

19. Røpke, I. (2004) The early history of modern ecological economics. Ecological Economics 50(3-4): 293-314. Røpke, I. (2005) Trends in the development of ecological economics from the late 1980s to the early 2000s. Ecological Economics 55(2): 262-290.

20. *Spash, C. L. (1999) The development of environmental thinking in economics. Environmental Values 8(4): 413-435. (PDF). Retrieved 2012-12-23.*

21. Costanza, R., Stern, D. I., He, L., Ma, C. (2004). Influential publications in ecological economics: a citation analysis. Ecological Economics 50(3-4): 261-292.

22. Spash, C. L. (2013) Influencing the perception of what and who is important in ecological economics. Ecological Economics 89: 204-209.

23. Spash, C. L. (2013) The Shallow or the Deep Ecological Economics Movement? Ecological Economics 93:351-362.

24. Anderson, B., M›Gonigle, M., 2012. Does ecological economics have a future?: contradiction and reinvention in the age of climate change. Ecological Economics 84, 37–48.

25. [g] Harris J. (2006). *Environmental and Natural Resource Economics: A Contemporary Approach*. Houghton Mifflin Company.

26. *Costanza R; et al. (1998). The value of the world›s ecosystem services and natural capitall. Ecological Economics.* ***25*** *(1): 3–15. doi:10.1016/S0921-8009(98)00020-2.*

27. Knapp G, Roheim CA and Anderson JL (2007) *The Great Salmon Run: Competition Between Wild And Farmed Salmon* World Wildlife Fund. ISBN 0-89164-175-0

28. Washington Post. Salmon Farming May Doom Wild Populations, Study Says.

29. Soil Association. UK Organic Group Exposes Myth that Genetically Engineered Crops Have Higher Yields. Organic Consumers Association.

30. Soderbaum P. (2004). Politics and Ideology in Ecological Economics. Internet Encyclopaedia of Ecological Economics.

31. http://www.clivespash.org/wp-content/uploads/2015/04/Spash_NEP_2009_EV.pdf

32. *Spash, C.L. (2011) Social ecological economics: Understanding the past to see the future. American Journal of Economics and Sociology 70, 340-375 (PDF). Retrieved 2014-01-07.*

33. *Jacqui Lagrue (2012-07-30). Spash, C.L., Ryan, A. (2012) Economic schools of thought on the environment: Investigating unity and division. Cambridge Journal of Economics 36, 1091-1121 (PDF). Cje.oxfordjournals.org. Retrieved 2014-01-07.*

34. Costanza R. (1989). What is ecological economics? *Ecological Economics* **1**:1-7. .

35. Martinez-Alier, J., 1994. Ecological economics and ecosocialism, in: O›Connor, M. (Ed.), Is Capitalism Sustainable? Guilford Press, New York, pp. 23-36

36. Spash, C.L., Clayton, A.M.H., 1997. The maintenance of natural capital: Motivations and methods, in: Light, A., Smith, J.M. (Eds.), Space, Place and Environmental Ethics. Rowman & Littlefield Publishers, Inc., Lanham, pp. 143-173

37. Toman, M., 1998. Why not to calculate the value of the world›s ecosystem services and natural capital. Ecological Economics 25, 57-60

38. O'Neill, J.F., 1993. Ecology, Policy and Politics: Human Well-Being and the Natural World. Routledge, London

39. O'Neill, J.F. (1997) Managing without prices: On the monetary valuation of biodiversity. Ambio 26, 546-550

40. Vatn, A., 2000. The environment as commodity. Environmental Values 9, 493-509

41. Norgaard, R. B. (1994) Development Betrayed: The End of Progress and a Coevolutionary Revisioning of the Future. London: Routledge

42. Daily, G.C. 1997. *Nature›s Services: Societal Dependence on Natural Ecosystems.* Washington, D.C.: Island Press.

43. Millennium Ecosystem Assessment. 2005. *Ecosystems and Human Well-Being: Biodiversity Synthesis*. Washington, D.C.: World Resources Institute.

44. McCauley, D. J. (2006) Selling out on nature. Nature 443(7): 27-28

45. Daly, H. and Farley, J. 2004. *Ecological Economics: Principles and Applications*. Washington: Island Press.

46. Ayres, R.U. 2007. On the practical limits of substitution. Ecological Economics 61: 115-128.

47. Shmelev, S.E. 2012. Ecological Economics. Sustainability in Practice, Springer

48. A. Müller, L. Kranzl, P. Tuominen, E. Boelman, M. Molinari, A.G. Entrop: Estimating exergy prices for energy carriers in heating systems: Country analyses of exergy substitution with capital expenditures. Energy and Buildings, Volume 43, Issue 12, December 2011, Pages 3609–3617.

49. *Hall, Charles A.S.; Cleveland, Cutler J.; Kaufmann, Robert (1992). Energy and Resource Quality: The ecology of the Economic Process. Niwot, Colorado: University Press of colorado.*

50. Peter A. Corning 1 *, Stephen J. Kline. (2000). Thermodynamics, information and life revisited, Part II: Thermoeconomics and Control information Systems Research and Behavioral Science, Apr. 07, Volume 15, Issue 6 , Pages 453 – 482

51. Corning, P. (2002). "Thermoeconomics – Beyond the Second Law" – source: www.complexsystems.org

52. *Wall, Göran. Exergy - a useful concept. Exergy.se. Retrieved 2012-12-23.*

53. *Environmental Decision making, Science and Technology. Telstar.ote.cmu.edu. Retrieved 2012-12-23.*

54. Stabile, Donald R. "Veblen and the Political Economy of the Engineer: the radical thinker and engineering leaders came to technocratic ideas at the same time," *American Journal of Economics and Sociology (45:1) 1986, 43-44.*

55. *Costanza, R.; d›Arge, R.; de Groot, R.; Farber, S.; Grasso, M.; Hannon, B.; Naeem, S.; Limburg, K.; Paruelo, J.; O›Neill, R.V.; Raskin, R.; Sutton, P.; and van den Belt, M. (1997). The value of the world›s ecosystem services and natural capital (PDF). Nature.* ***387*** *(6630): 253–260. Bibcode:1997Natur.387..253C. doi:10.1038/387253a0.*

56. *Norgaard, R.B.; Bode, C. (1998). Next, the value of God, and other reactions. Ecological Economics.* ***25****: 37–39. doi:10.1016/s0921-8009(98)00012-3.*

57. *Brouwer, Roy (January 2000). Environmental value transfer: state of the art and future prospects. Ecological Economics.* ***32*** *(1). doi:10.1016/S0921-8009(99)00070-1. Retrieved 11 November 2016.*

58. *Gómez-Baggethun, Erik; de Groot, Rudolf; Lomas, Pedro; Montes, Carlos (1 April 2010). The history of ecosystem services in economic theory and practice: From early notions to markets and payment schemes. Ecological Economics.* ***69*** *(6). doi:10.1016/j.ecolecon.2009.11.007. Retrieved 11 November 2016.*

59. *Farber, Stephen; Constanza, Robert; Wilson, Matthew (June 2002). Economic and ecological concepts for valuing ecosystem services. Ecological Economics.* ***41*** *(3). doi:10.1016/S0921-8009(02)00088-5. Retrieved 11 November 2016.*

60. SocialEdge.org. Accessed: December 23, 2012.

61. Archived June 21, 2008, at the Wayback Machine.

62. Multinational Monitor, 9/2007. Accessed: December 23, 2012.

63. *Ecuador threat to drill jungle oil. Archived from the original on December 18, 2008.*

64. *International News | World News - ABC News. Abcnews.go.com. 4 June 2012. Retrieved 2012-12-23.*

65. http://www.panna.org/jt/agAssessment

66. *Emmott, Bill (April 17, 2008). GM crops can save us from food shortages. The Daily Telegraph. London.*

67. *Costanza, Robert; Segura,Olman; Olsen, Juan Martinez-Alier (1996). Getting Down to Earth: Practical Applications of Ecological Economics. Washington, D.C.: Island Press. ISBN 1559635037.*

68. Pearce, Fred "Blueprint for a Greener Economy"

69. *Spash, C. L. (2007) The economics of climate change impacts à la Stern: Novel and nuanced or rhetorically restricted? Ecological Economics 63(4): 706-713 (PDF). Retrieved 2012-12-23.*

70. Hawken, Paul (1994) "The Ecology of Commerce" (Collins)

71. Hawken, Paul; Amory and Hunter Lovins (2000) Natural Capitalism: Creating the Next Industrial Revolution (Back Bay Books)

72. Martinez-Alier, Joan (2002) The Environmentalism of the Poor: A Study of Ecological Conflicts and Valuation. Cheltenham, Edward Elgar

73. Kapp, Karl William (1963) The Social Costs of Business Enterprise. Bombay/ London, Asia Publishing House.

74. Kapp, Karl William (1971) Social costs, neo-classical economics and environmental planning. The Social Costs of Business Enterprise, 3rd edition. K. W. Kapp. Nottingham, Spokesman: 305-318

75. Einsentein, Charles (2011), Sacred Economics: Money, Gift and Society in an Age in Transition (Evolver Editions)

76. *Spash, Clive L. (16 July 2010). The brave new world of carbon trading. New Political Economy. Taylor and Francis Online.* ***15*** *(2): 169–195. doi:10.1080/13563460903556049.* Copy also available at http://www.clivespash.org/2010_Spash_Brave_New_World_NPE.pdf

77. Proops, J. , and Safonov, P. (eds.) (2004), Modelling in Ecological Economics, Edward Elgar

78. Faucheux, S., Pearce, D., and Proops, J. (eds.) (1995), Models of Sustainable Development, Edward Elgar

79. *Chen, Jing (2015). The Unity of Science and Economics: A New Foundation of Economic Theory. http://www.springer.com/us/book/9781493934645: Springer.*

80. Costanza, R., and Voinov, A. (eds.) (2004), Landscape Simulation Modeling. A Spatially Explicit, Dynamic Approach, Springer-Verlag New-York, Inc.

81. *Voinov, Alexey (2008). Systems science and modeling for ecological economics (1st ed.). Amsterdam: Elsevier Academic Press. ISBN 0080886175.*

82. Felber, Christian (2012), La economia del bien commun (Duestro)

83. Mace GM. Whose conservation? Science (80-). 2014 Sep 25;345(6204):1558–60.

84. Dasgupta P. Nature's role in sustaining economic development. Philos Trans R Soc Lond B Biol Sci. 2010 Jan 12;365(1537):5–11.

85. McCauley D. Selling out on nature. Nature. 2006;443(September):7–8.

86. Mech LD. The Challenge and Opportunity of Recovering Wolf Populations. Conserv Biol. 1995 Apr;9(2):270–8

87. Ricketts TH, Daily GC, Ehrlich PR, Michener CD. Economic value of tropical forest to coffee production. Proc Natl Acad Sci U S A. 2004 Aug 24;101(34):12579–82

CHAPTER 4

Balance of Nature

Chapter Contents

To begin to understand the balance of nature, a few concepts we have not yet considered are useful. However, we will start with a question. In the previous submodule on growth we saw that organisms are endowed with the capacity for exponential growth. How can nature achieve balance if each living organism is capable of taking over the environment? In this respect, balance and stability are similar concepts. Some kind of restraint is necessary to orchestrate the balance among all the different organism making up the ecosystem. We considered negative feedbacks, a general concept about restraint, without going into more detail. Certainly negative feedbacks play an important role in achieving balance, but an evolved "self-restraint"is also part of the balance of nature. As you might expect, this simple idea of the balance of nature, is a very complex process. We look at some aspects of this complexity below.

INTRODUCTION

The balance of nature is a theory that proposes that ecological systems are usually in a stable equilibrium (homeostasis), which is to say that a small change in some particular parameter (the size of a particular population, for example) will be corrected by some negative feedback that will bring the parameter back to its original «point of balance» with the rest of the system. It may apply where populations depend on each other, for example in predator/prey systems, or relationships between herbivores and their food source. It is also sometimes applied to the relationship between the Earth›s ecosystem, the composition of the atmosphere, and the world›s weather.

The Gaia hypothesis is a balance of nature-based theory that suggests that the Earth and its ecology may act as co-ordinated systems in order to maintain the balance of nature.

The theory that nature is permanently in balance has been largely discredited, as it has been found that chaotic changes in population levels are common, but nevertheless the idea continues to be popular. During the later half of the twentieth century the theory was superseded by catastrophe theory and chaos theory.

HISTORY OF THE THEORY

The concept that nature maintains its condition is of ancient provenance; Herodotus commented on the wonderful relationship between predator and prey species, which remained in a steady proportion to one another, with predators never excessively consuming their prey populations. The "balance of nature" concept once ruled ecological research, as well as once governing the management of natural resources. This led to a doctrine popular among some conservationists that nature was best left to its own devices, and that human intervention into it was by definition unacceptable.

PREDATOR PREY INTERACTIONS

Predator-prey populations tend to show chaotic behavior within limits, where the sizes of populations change in a way that may appear random, but is in fact obeying deterministic laws based only on the relationship between a population and its food source illustrated by the Lotka–Volterra equation. An experimental example of this was shown in an eight-year study on small Baltic Sea creatures such as plankton, which were isolated from the rest of the ocean. Each member of the food web was shown to take turns multiplying and declining, even though the scientists kept the outside conditions constant. An article in *Journal Nature* stated; "Advanced mathematical techniques proved the indisputable presence of chaos in this food web ... short-term prediction is possible, but long-term prediction is not."

HUMAN INTERVENTION

Although some conservationist organizations argue that human activity is incompatible with a balanced ecosystem, there are numerous examples in history showing that several modern day habitats originate from human activity: some of Latin America's rain forests owe their existence to humans planting and transplanting them, while the abundance of grazing animals in the Serengeti plain of Africa is thought by some ecologists to be partly due to human-set fires that created savanna habitats.

Possibly one of the best examples of an ecosystem fundamentally modified by human activity can be observed as a consequence of the Australian Aboriginal practice of "Fire-stick farming". The legacy of this practice over long periods has resulted in forests being converted to grasslands capable of sustaining larger populations of faunal prey, particularly in the northern and western regions of the continent. So thorough has been the effect of these deliberate regular burnings that many plant and tree species from affected regions have now completely adapted to the annual fire regime in that they require the passage of a fire before their seeds will even germinate.

CONTINUED POPULARITY OF THE THEORY

Despite being discredited among ecologists, the theory is widely held to be true in the wider population: a report written by psychologist Corinne Zimmerman of Illinois State University and ecologist Kim Cuddington of Ohio University demonstrated that at least in Midwestern America, the "balance of nature" idea is widely held among both science majors and the general student population. Moreover, in a study by Ergazaki & Ampatzidis (2012), educational sciences students were asked to reason about the future of ecosystems which suffered human-driven disturbances. The researchers report that their informants found it very likely for the ecosystems to fully recover their initial state. To justify their predictions, they referred to either a 'recovery process' which restores the initial 'balance', or specific 'recovery mechanisms' as an ecosystems' inherent

BIOLOGICAL DATA VISUALIZATION

Biology Data Visualization is a branch of bioinformatics concerned with the application of computer graphics, scientific visualization, and information visualization to different areas of the life sciences. This includes visualization of sequences, genomes, alignments, phylogenies, macromolecular structures, systems biology, microscopy, and magnetic resonance imaging data. Software tools used for visualizing biological data range from simple, standalone programs to complex, integrated systems.

State-of-the-art and Perspectives

Today we are experiencing a rapid growth in volume and diversity of biological data, presenting an increasing challenge for biologists. A key step in understanding and learning from these data is visualization. Thus, there has been a corresponding increase in the number and diversity of systems for visualizing biological data.

An emerging trend is the blurring of boundaries between the visualization of 3D structures at atomic resolution, visualization of larger complexes by cryo-electron microscopy, and visualization of the location of proteins and complexes within whole cells and tissues.

A second emerging trend is an increase in the availability and importance of time-resolved data from systems biology, electron microscopy and cell and tissue imaging. In contrast, visualization of trajectories has long been a prominent part of molecular dynamics.

Finally, as datasets are increasing in size, complexity, and interconnectness, biological visualization systems are improving in usability, data integration and standardization.

List of Visualization Systems

A large number of software systems are available for visualization biological data. The links below link lists of such systems, grouped by application areas.

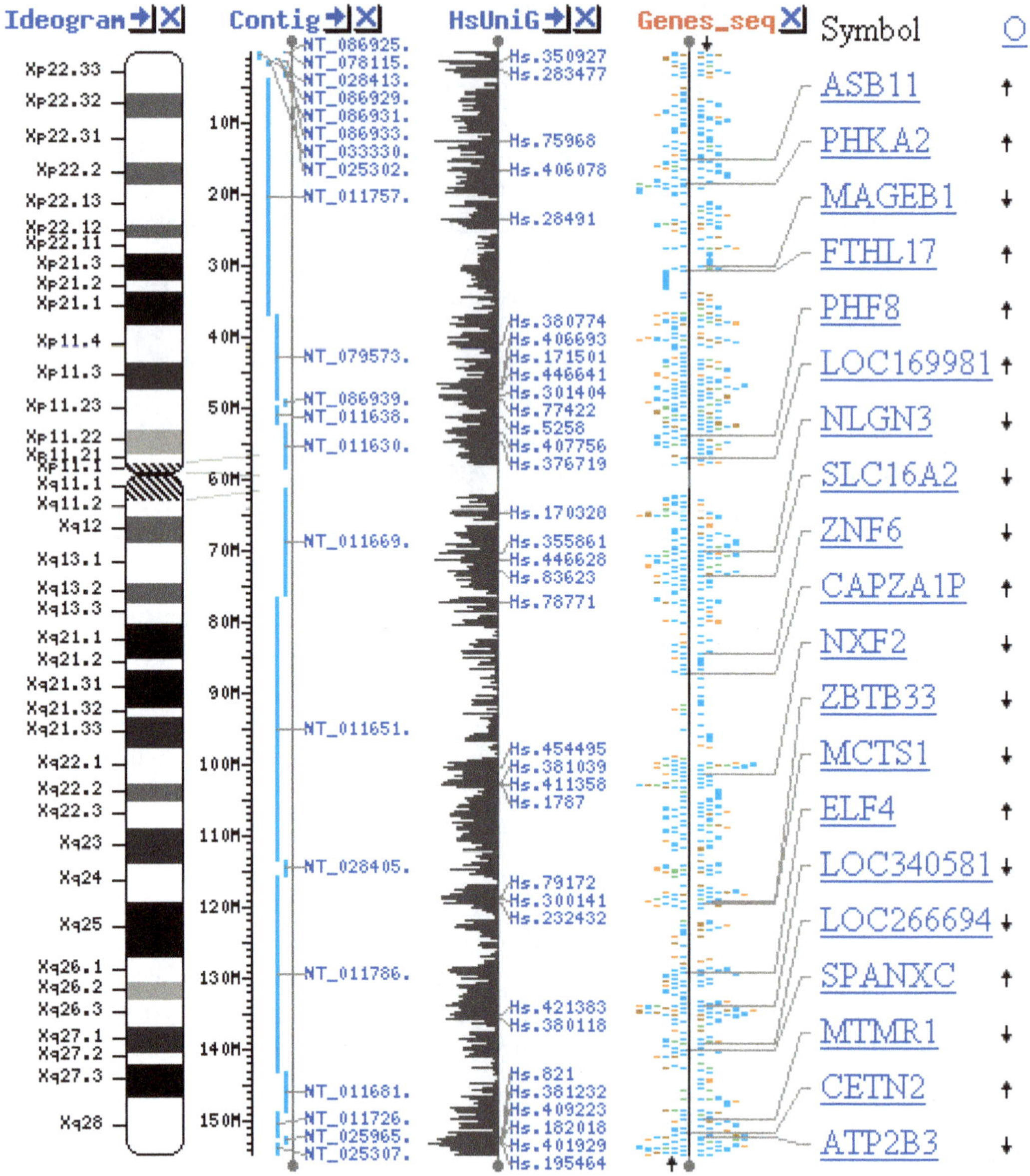

Genomic and assembly data

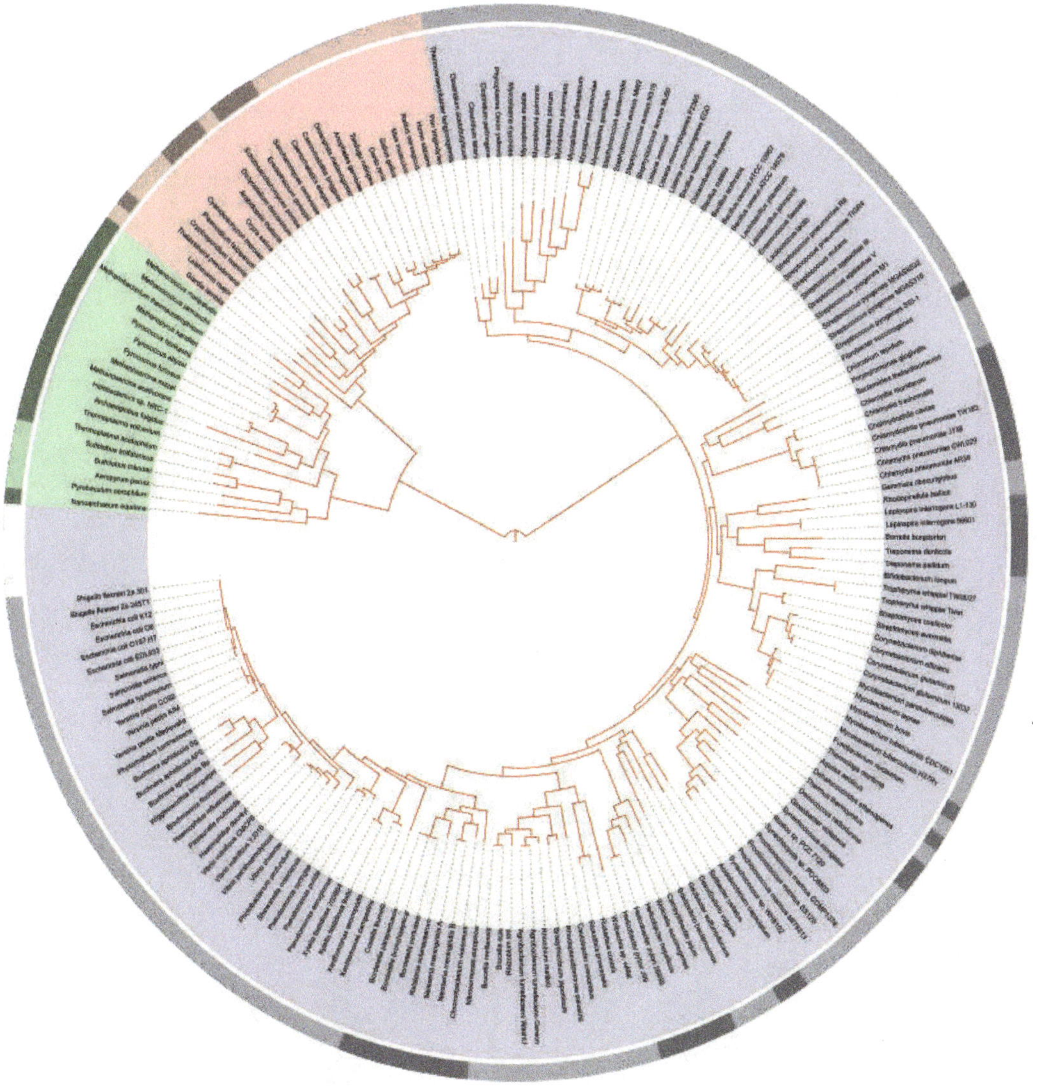

Alignments, phylogeny and evolution

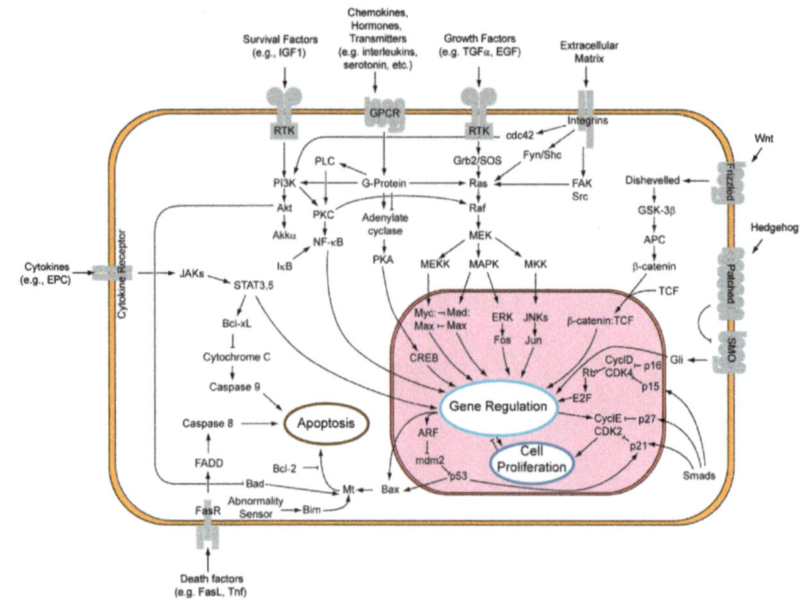

Systems biology

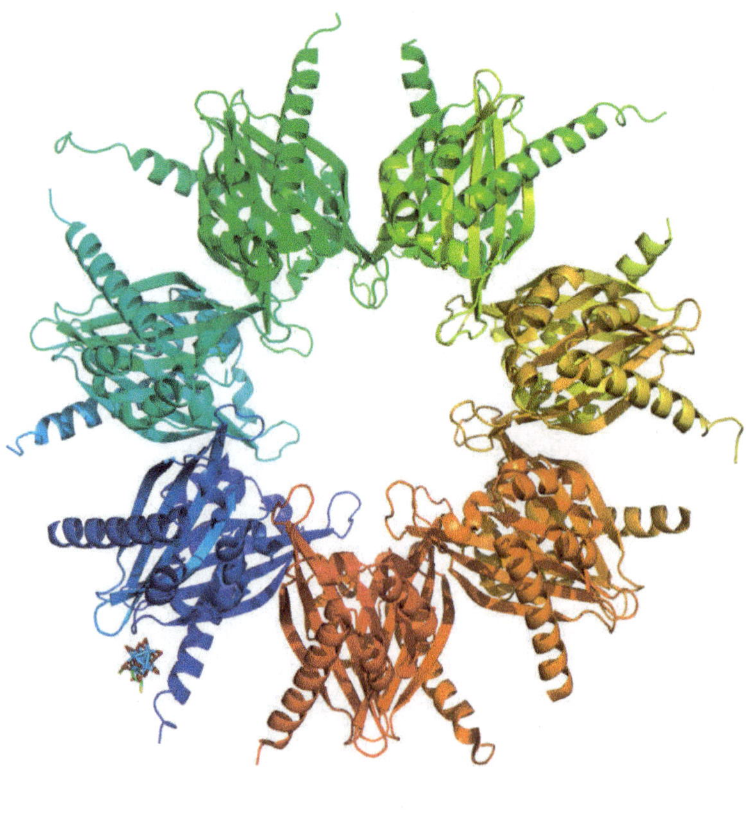

Molecular GraphicsMolecular Dynamics

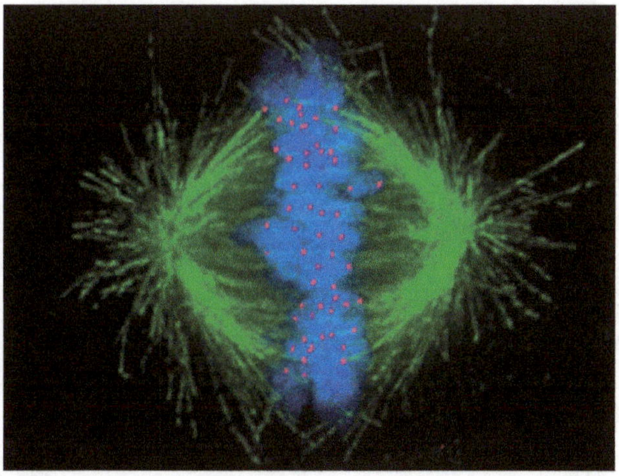

Microscopy

Magnetic resonance imaging

REFERENCES

1. Zimmerman, Corinne (October 2007). "Ambiguous, circular and polysemous: students' definitions of the "balance of nature" metaphor". Public Understanding of Science. 16 *(4): 393–406.* doi:10.1177/0963662505063022. Retrieved 19 June 2011.
2. Jacobs, Tom (27 November 2007). «Belief in ‹Balance of Nature› Hard to hi». Miller-McCune. Archived from the original on 29 July 2012. Retrieved 19 June 2011.
3. Stevens, William K. (31 July 1990). «New Eye on Nature: The Real Constant Is Eternal Turmoil». The New York Times. Retrieved 19 June 2011.
4. The Ottawa Citizen (13 February 2008). «Study of ocean life shows a «chaotic» balance of nature». CanWest MediaWorks Publications Inc. Archived from the original on 24 June 2010. Retrieved 20 June 2011.
5. Ergazaki, M., & Ampatzidis, G. (2012). Students' Reasoning about the Future of Disturbed or Protected Ecosystems & the Idea of the 'Balance of Nature'. Research in Science Education, 42(3) 511-530.
6. Lucić (2005). «Structural studies by electron tomography: from cells to molecules». Annual Review of Biochemistry. 74: 833–65. doi:10.1146/annurev.biochem.73.011303.074112. PMID 15952904. |first2= missing |last2= in Authors list (help); |first3=missing |last3= in Authors list (help)
7. Steven (2008). «The future is hybrid». Journal of structural biology. 163 *(3): 186–95. doi:10.1016/j.jsb.2008.06.002. PMID 18602011.* |first2= missing |last2= in Authors list (help)
8. Plattner (2006). «Sub-second cellular dynamics: time-resolved electron microscopy and functional correlation». International review of cytology. International Review of Cytology. 255: 133–76. doi:10.1016/S0074-7696(06)55003-X. ISBN 9780123735997. PMID 17178466. |first2= missing |last2= in Authors list (help)
9. Frank (2004). «Time-resolved imaging of macromolecular processes and interactions». Journal of structural biology. 147 *(3): 209–10. doi:10.1016/j.jsb.2004.06.003. PMID 15450290.* |first2= missing |last2= in Authors list (help)

CHAPTER 5

Constructal Law

Chapter Contents

- **Introduction**
- **Manifestations**
- **Evolutionary Design**
- **Constructal Thermodynamics**
- **Criticisms**
- **Cline Biology**

Constructal law is a theory of the generation of design (configurations, patterns, geometry) in nature. According to this theory, natural design and the constructal law unite all animate and inanimate systems. The constructal law was stated by Adrian Bejan in 1996 as follows: "For a finite-size system to persist in time (to live), it must evolve in such a way that it provides easier access to the imposed currents that flow through it." "Constructal" is a word coined by Bejan to describe the natural tendency of flow systems (e.g. rivers, trees and branches, lungs, tectonic plates and engineered forms) to generate and evolve structures that increase flow access.

INTRODUCTION

The constructal law was proposed in 1996 as a summary of all design generation and evolution phenomena in nature, bio and non-bio. The constructal law represents three steps toward making "design in nature" a concept and law-based domain in science:

- Life is flow: all flow systems are living systems, the animate and the inanimate.
- Design generation and evolution is a phenomenon of physics.
- Designs have the universal tendency to evolve in a certain direction in time.

The constructal law is proposed as a first principle of physics accounting for all design and evolution in nature. It holds that shape and structure arise to facilitate flow. The designs that happen spontaneously in nature reflect this tendency: they allow entities to flow more easily – to measurably move more current farther and faster per unit of useful energy consumed. Rain drops, for example, coalesce and move together, generating rivulets, streams and the mighty river basins of the world because this design allows them to move more easily.

The constructal law asks the question: Why does this design arise at all? Why can't the water just seep through the ground? The constructal law provides this answer: Because the water flows better with design. The constructal law covers the tendency of nature to generate designs to facilitate flow.

MANIFESTATIONS

The constructal law covers natural phenomena of organization, such as tree-shaped flows, round tubes and bones, scaling laws, etc. The lightning bolts that flash across the sky generate a tree-like structure because this is a good design for moving a current (electricity) from an area (the cloud) to a point (a church steeple or another cloud). The circulatory and nervous systems of biological creatures generate a similar tree-like design because they too are moving currents from a point to an area and from an area to a point.

Although treelike structures are a very common design in nature, they are only one manifestation of the constructal law. In a simple example, logs floating on a lake or icebergs at sea orient themselves perpendicular to the wind which increases the transfer of motion from the moving air body to the water body. A more complex example is the design of animals that have evolved to move mass more efficiently (to cover more distance per unit of useful energy) across the landscape.

This includes the seemingly "characteristic" sizes of organs, the shape of bones, the rhythm of breathing lungs and beating hearts, of undulating tails, running legs, and flapping wings. The constructal law proclaims that all these designs have arisen—and work together—to allow animals, like raindrops in a river basin, to move more easily across a landscape. Because human beings are not separate from but a part of nature, their designs are also governed by the constructal law.

EVOLUTIONARY DESIGN

The constructal law defines the time direction of all evolutionary design phenomena. It states that designs should evolve over time to acquire better configurations to provide more access for the currents that flow through them. It defines in physics terms what it means to be "better", more "fit", to "survive", and to be efficient. Not all changes are improvements, but those that stick measurably enhance flow.

Constructal design occurs at every scale. Each component of an evolving flow system—each rivulet, each tree, and each road—acquires evolving designs to facilitate flow access. As these elements coalesce into larger and larger structures (into evolving river basins, forests and transport networks), a hierarchy emerges such that the multi sized components and channels work together so that everything flows more easily. This is seen in the shape and structure of the neural networks in the brain, of the alveoli in the lung, the size and distribution of vegetation in the forest and of human settlements on the map.

In the big picture, all the mating and morphing flows on the largest system that surrounds us, the Earth itself, evolve to enhance global flow. For example, trees and other forms of vegetation that move moisture from the ground to the air are components of the larger global system, including forests, river basins and weather patterns, that have the tendency to equilibrate all the moisture on Earth. The constructal law states that every flow system is destined to remain imperfect. The direction of design evolution is toward distributing the imperfections of the system, such that the "whole" flows easier (e.g., river basin, animal body, human vehicle).

Evolution never ends. Optimality statements (minimum, maximum, optimum, static, end design, destiny) have only local, limited applicability. The constructal law covers them because it is about the time direction of all the evolutionary design phenomenon.

The constructal law is a law of physics - the law of design generation and evolution in nature. The natural phenomenon is not the elimination but the distribution (better and better over time) of imperfection. The distribution of imperfection generates the geometry (shape, structure) of the system. Today, optimization of many systems arising in thermal engineering such as conductive pathways (fins, cavities and highly conductive inserts), convective channels and radiative systems are inspired by constructal law.

For example, in point-area and point-volume flows, the constructal law predicts tree architectures, such flows displaying at least two regimes: one highly resistive and one with lower resistivity. The constructal-law tendency manifests itself at every scale.

The tree is the natural flow design for achieving flow access between one point and a volume. Alternating trees achieve flow access between two planes. Natural porous media exhibit multi-scale flow structures consistent with the multiple scales and performance of alternating trees.

Some domains of application

Application	What flows	Tree channels: Low Resistivity	Interstitial spaces: High Resistivity
Packages of electronics	Heat	High-conductivity inserts (blades, needles)	Low conductivity substrate
Urban traffic	People	Low-resistance street car traffic	Street walking in urban structure
River basins	Water	Low-resistance rivulet and rivers	Darcy flow through porous media
Lungs	Air	Low-resistance airways, bronchial passages	diffusion in alveoli tissues

Circulatory system	Blood	Low-resistance blood vessels, capillaries, arteries, veins	diffusion in capillaries tissues

The constructal law provides a unifying theory of evolution. It holds that inanimate and animate phenomena generate evolving configurations to move more easily. The constructal law also provides the physics definition of life, of what it means to be alive. It states that life means flow and the free generation of design. If the flows stop, the system is dead (in thermodynamic equilibrium). The constructal law is the physics law of life, design and evolution.

CONSTRUCTAL THERMODYNAMICS

Thermodynamics rests on two laws. Both are first principles: The first law commands the conservation of energy, and the second law commands irreversibility: the tendency of all currents to flow from high (temperature, pressure) to low. These two laws are about systems in the most general sense, viewed as black boxes, without shape and structure.

The two laws of thermodynamics do not account for nature completely. Nature is not made of black boxes. Nature's boxes are filled with evolving, freely morphing configurations—even the fact that they have names (rivers, blood vessels) is due to their appearance, organization, or design. Where the second law commands that things should flow from high to low, the constructal law commands that they evolve in configurations that flow more and more easily over time.

Classical thermodynamics versus constructal thermodynamics

Thermodynamics	**Constructal theory**
State	Flow architecture (flow structure)
Process	Change of structure (design change)
Properties	Global objective and global constraints
Equilibrium state	Equilibrium flow architecture
Fundamental relation	Fundamental relation
Constrained equilibrium states	Nonequilibrium architectures
Removal of constraints	Increased freedom to morph
Energy minimum principle	Evolution toward greater flow access

In contrast to fractal models of observed objects in nature, the constructal law is predictive and thus can be tested experimentally. Many natural designs, animate and inanimate, have been explained and unified by the constructal law. For example:

Global convective cells on Earth's atmosphere	River basin design: Horton's rules of stream numbers (~4) and lengths (~2), and all the other scaling rules (e.g., Melton, Hack) of river basins all over the world
The distribution of city sizes and numbers, i.e. Zipf's law relating log (size) versus log (rank)	The distribution of tree sizes and numbers on the forest floor
Vision, cognition, and the golden ratio phenomenon	Pedestrian movement, speeds, and patterns
The flow of education as a morphing vasculature on the globe, and the rigidity of university rankings	The entire architecture of vegetation: roots, trunks, canopies, branches, leaves, and the forest, including the prediction of Leonardo da Vinci's rule, Huber's rule, and the Fibonacci sequence.
The emergence of urban traffic design.	The entire morphogenesis of dendritic crystals (e.g., snowflakes), as a flow structure that facilitates the flow of the heat of solidification.
The scaling law of all animal locomotion (running, flying, swimming): speeds, frequencies, forces and the work spent per unit of mass moved and distance traveled.	The evolution of speed in sports.
Kleiber's law, the relationship between metabolic rate and body size.	The relationship between breathing and heart beating times and body size.
The human bronchial tree with 23 levels of bifurcation.	The relationship between the mass transfer contact area and body mass.
The dimensions of the alveolar sac.	The total length of the airways.
The total alveolar surface area.	The total resistance to oxygen transport in the respiratory tree.
The life time and life travel of animals, vehicles (human & machine species), rivers and air currents.	Aspect ratios of airplanes

CRITICISMS

One reservation toward the constructal law is that its formulation is vague. The constructal law states that "For a finite-size system to persist in time (to live), it must evolve in such a way that it provides easier access to the imposed currents that flow through it", but there is neither a mention of what these "currents" are nor an explicit definition of what "providing easier access" means, nor precisely formulating the relationship with math. Without defining the physical quantities or their exact relationships, it is not physical or mathematical. There has been no attempt to prove it from first principles. Contrarily to alternative theories of non-equilibrium thermodynamics.

There is also the claim that "the constructal law can be seen as a complement to MEP (maximum entropy production)".

Responses To Criticisms

No criticism has been published in the peer reviewed literature against the constructal law as a law of physicsStrictly speaking refs. published after 1996 do not refer to the constructal law at all. There is one scholar who questioned the clarity of the language used in the constructal law. In ref., Kleidon wrote "It is as yet unclear how the term 'access' in the definition of the constructal theory (sic) can be quantified, but its original roots are undoubtedly in thermodynamics". Bejan has responded by noting that one cannot prove a first principle based on other first principles. The constructal law is not about what flows, but about the physics phenomenon of how any flow system acquires its evolving configuration (design) over time. The constructal law is not about optimality (max, min, opt)—it is the definition of "life" in physics terms, and of the time direction of the changes in flow configuration.

Bejan also noted that the phenomenon governed by the constructal law (design in nature) is macroscopic (finite size, not infinitesimal). It is the birth of design and evolution of design in all the parts together. It is dynamic, not static. The evolution never ends. There is no end design, no destiny (max, min, opt).

Bejan and Lorente expanded on this by explaining the difference between a law of physics (e.g., the constructal law) and the many invocations of the law, which underpin many "theories" based on the law. In the section titled "The constructal law versus the second law of thermodynamics" they noted that the constructal law and the second law are first principles. The constructal law is a useful reminder not only of what was missing in physics and thermodynamics (the law of design and evolution) but also of what is present. For example, contrary to the critic's view, the first and second laws of thermodynamics did not require any "proof based on simplified systems of statistical physics."

The constructal law is a statement of a natural tendency—the time direction of the phenomenon. It is as non-mathematical as the original statements of the second law:

Clausius: No process is possible whose sole result is the transfer of heat from a body of lower temperature to a body of higher temperature.

Kelvin: Spontaneously, heat cannot flow from cold regions to hot regions without external work being performed on the system.

A new law does not have to be stated in mathematical terms. The mathematization of the second law statement (and of thermodynamics) came later. The constructal law underwent the same evolution. The 1996 statement was followed in 2004 by a complete mathematical formulation of constructal-law thermodynamics.

Flow leads to better flow. Like any other law of physics, the constructal law is a concise summary of observed facts: the natural tendency of flow systems to evolve toward configurations that provide easier access over time. The word "access" means the ability to move through a confined space such as a crowded room. This is why "finite size" appears in the constructal law statement. This mental viewing covers all the flow design and evolution phenomena, animate and inanimate, because they all morph to enter and to flow better, more easily.

Finally, Reis has showed that MEP (maximum entropy production), like its complete opposite (minimum entropy production) is a special ad-hoc optimality principle that is covered by the constructal law.

CLINE (BIOLOGY)

In biology and ecology, an ecocline or simply cline (from Greek: κλίνω "to possess or exhibit gradient, to lean") describes an ecotone in which a series of biocommunities display a continuous gradient. The term was coined by the English evolutionary biologist Julian Huxley in 1938.

More technically, clines consist of ecotypes or forms of species that exhibit gradual phenotypic and/or genetic differences over a geographical area, typically as a result of environmental heterogeneity. Genetically, clines result from the change of allele frequencies within the gene pool of the group of taxa in question. Clines may manifest in time and/or space.

Gradient Analysis

In ecology, spatial clines have led to gradient analysis where the abundance and distribution of organisms is rendered by sinusoidal curves on the plane. From these curves can be extracted that populations occupy zones of maximum and minimum presence, according to their special needs and tolerances imposed by their environment.

Typically, a well-marked cline does not allow for a delineation of subspecies, as it is then impossible, by definition, to draw any further clear dividing lines between populations. In population genetics, a cline could include a spectrum of subspecies, as allele and haplotype frequencies tend to vary over a larger space; moreover, in evolution, genetic lineage sorting usually lags behind the establishment of locally differentiated phenotypes. Regardless, in neither case will such a variation yield different species, as long as the populations, though geographically spread, can interbreed one with another.

Ring Species

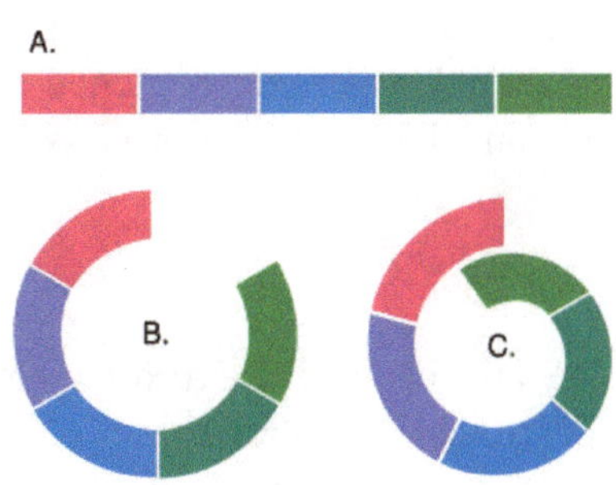

Interbreeding populations represented by coloured blocks

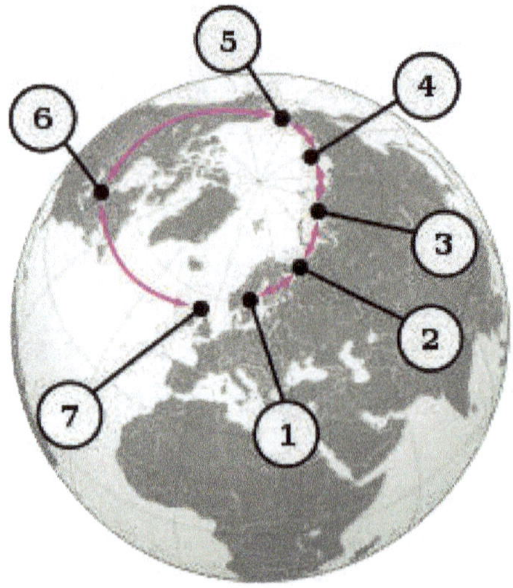

The *Larus* gulls interbreed in a ring around the arctic.

1: *Larus argentatus argentatus*, 2: *Larus fuscus* (*sensu stricto*), 3: *Larus fuscus heuglini*, 4: *Larus argentatus birulai*, 5: *Larus argentatus vegae*, 6: *Larus argentatus smithsonianus*, 7: *Larus argentatus argenteus*

Ring species are a distinct type of cline where the geographical distribution in question is circular in shape, so that the two ends of the cline overlap with one another, giving two adjacent populations that rarely interbreed due to the cumulative effect of the many changes in phenotype along the cline. The populations elsewhere along the cline interbreed with their geographically adjacent populations as in a standard cline. Ring species present an interesting problem for those who seek to divide the living world into discrete species, as chain species are closely related to speciation (in this case, parapatric).

In the case of *Larus* gulls, the habitats of the end populations even overlap, which introduces questions as to what constitutes a species: nowhere along the cline can a line be drawn between the populations, but they are unable to interbreed. However a recent study (Liebers et al., 2004) has provided genetic evidence that this example is far more complicated than presented here, and likely does not constitute a typical ring species.

REFERENCES

1. T. Basak (2011), "The law of life: the bridge between physics and biology: comment on "The constructal law and the evolution of design in nature" by A. Bejan and S. Lorente", Phys Life Rev 8(3), pp. 249–52 (doi: 10.1016/j.plrev.2011.07.003); A. Bejan, S. Lorente (2010), "The constructal law and the evolution of design in nature", Philosophical Transactions B vol. 365, issue 1545 (doi: 10.1098/rstb.2009.0302).
2. Bejan (1997), Advanced Engineering Thermodynamics (2nd ed.), New York: Wiley.
3. Kremer-Marietti, J. Dhombres (2006), L'Épistemologie, Paris: Ellipses.

4. Tondeur, D.; Fan, Y.; Luo, L. (2009). "Constructal optimization of arborescent structures with flow singularities". Chem Eng Sci. 64: 3968–3982. doi:10.1016/j.ces.2009.05.052.

5. Quéré, S. (2010). "Constructal theory of plate tectonics". Int J Design & Nature Ecodyn. 5: 242–253. doi:10.2495/dne-v5-n3-242-253.

6. D. Queiros-Conde, M. Feidt (eds., 2009), Constructal Theory and Multi-scale Geometries: Theory and Applications in Energetics, Chemical Engineering and Materials, Paris: Les Presses de L'ENSTA.

7. Reis, A. H. (2006). "Constructal theory: from engineering to physics, and how flow systems develop shape and structure". Appl Mech Rev. 59: 269–282. Bibcode:2006ApMRv..59..269R. doi:10.1115/1.2204075.

8. Wang, L. (2011). "Universality of design and its evolution". Phys Life Rev. 8: 257–258. Bibcode:2011PhLRv...8..257W. doi:10.1016/j.plrev.2011.08.003.

9. N. Acuña, Mindshare. Igniting Creativity and Innovation Through Design Intelligence (Motion, Henderson, Nevada 2012).

10. Miguel, A. F. (2011). "The physics principle of the generation of flow configuration". Phys Life Rev. 8: 243–244. Bibcode:2011PhLRv...8..243M. doi:10.1016/j.plrev.2011.07.006.

11. Hajmohammadi, MR; Eskandari, H; Saffar-Avval, M; Campo, A (2013). "A new configuration of bend tubes for compound optimization of heat and fluid flow". Energy. 62: 418–424. doi:10.1016/j.energy.2013.09.046.

12. Hajmohammadi, M. R.; Poozesh, S.; Campo, A.; Salman Nourazar, Seyed (2013). "Valuable reconsideration in the constructal design of cavities". Energy Conversion and Management. 66: 33–40. doi:10.1016/j.enconman.2012.09.031.

13. Miguel, A. F. (2006). "Constructal pattern formation in stony corals, bacterial colonies and plant roots under different hydrodynamics conditions". Journal of Theoretical Biology. 242: 954–961. doi:10.1016/j.jtbi.2006.05.010.

14. Miguel, A. F.; Bejan, A. (2009). "The principle that generates dissimilar patterns inside aggregates of organisms". Journal Physica A. 388: 727–731. Bibcode:2009PhyA..388..727M. doi:10.1016/j.physa.2008.11.013.

15. H. Reis and C. Gama, Sand size versus beachface slope – an explanation based on the Constructal Law, Geomorphology 114, 276-283 (2010).

16. Reis, A. H. (2006). "Constructal view of scaling laws of river basins". Geomorphology. 78: 201–206. Bibcode:2006Geomo..78..201R. doi:10.1016/j.geomorph.2006.01.015.

17. Miguel, A. F. (2010). "Natural flow systems: acquiring their constructal morphology". Int J Design & Nature Ecodyn. 5: 230–241. doi:10.2495/dne-v5-n3-230-241.

18. Reis, A. H. (2011). "Design in nature, and the laws of physics". Phys Life Rev. 8: 255–256. Bibcode:2011PhLRv...8..255R. doi:10.1016/j.plrev.2011.07.001.

19. Bejan, A; Marden, James H. (2006). Constructing Animal Locomotion from New Thermodynamics Theory. American Scientist, July–August, Volume 94, Number 4

20. G. Resconi, Morphotronics and Constructal theory, LINDI 2011, 3rd IEEE Int. Symp. Logistics and Industrial Informatics, Budapest, Hungary, August 25–27 (2011).

21. L. C. Kelley and K. Behan, Empathy & evolution: how dogs convert stress into flow. Psychology Today 6 August 2012.

22. L. C. Kelley and K. Behan, The canine mind bows to the Constructal Law. Psychology Today 16 October (2012).

23. Bejan, A (2010). "The Constructal Law Origin of the Wheel, Size, and Skeleton in Animal Design". American Journal of Physics. 78 (7): 692–699. Bibcode:2010AmJPh..78..692B. doi:10.1119/1.3431988.

24. Miguel, Antonio F. (27 April 2009). "Constructal theory of pedestrian dynamics". Physics Letters A. 373 (20): 1734–1738. Bibcode:2009PhLA..373.1734M. doi:10.1016/j.physleta.2009.03.020.

25. Bejan, A; Merkx, Gilbert A, eds. (2007). "Constructal Theory of Social Dynamics." New York: Springer.

26. P. Kalason, Épistémologie Constructale du Lien Cultuel (L'Harmattan, Paris, 2007).

27. Bejan, Adrian (2005). "The Constructal Law of Organization in Nature: Tree-shaped flows and body size". Journal of Experimental Biology. 208 (9): 1677–1686. doi:10.1242/jeb.01487.

28. Lorente, S; Bejan, A (2010). "Few Large and Many Small: Hierarchy in Movement on Earth". International Journal of Design & Nature and Ecodynamics. 5 (3): 1–14.

29. Bejan, Adrian; Lorente, Sylvie; Lee, J. (7 October 2008). "Unifying constructal theory of tree roots, canopies and forests". Journal of Theoretical Biology. 254 (3): 529–540. doi:10.1016/j.jtbi.2008.06.026.

30. Ventikos, Y. (2011). "The importance of the constructal framework in understanding and eventually replicating structure in tissue". Phys Life Rev. 8: 241–242. Bibcode:2011PhLRv...8..241V. doi:10.1016/j.plrev.2011.07.007.

31. Bejan, A; Lorente, S (2008). "Design with Constructal Theory," Hoboken: Wiley.

32. Hajmohammadi, M. R.; Shirani, E.; Salimpour, M. R.; Campo, A. (2012). "Constructal placement of unequal heat sources on a plate cooled by laminar forced convection". International Journal of Thermal sciences. 60: 13–22. doi:10.1016/j.ijthermalsci.2012.04.025.

33. Hajmohammadi, M. R.; Poozesh, S.; Nourazar, S. S. (2012). "Constructal design of multiple heat sources in a square-shaped fin". Journal of Process Mechanichal Engineering. 226: 324–336. doi:10.1177/0954408912447720.

34. Hajmohammadi, M. R.; Abianeh, V. Alizadeh; Moezzinajafabadi, M.; Daneshi, M. (2013). "Fork-shaped highly conductive pathways for maximum cooling in a heat generating piece". Applied Thermal Engineering. 61: 228–235. doi:10.1016/j.applthermaleng.2013.08.001.

35. Eslami, M.; Jafarpur, K. (2012). "Optimal distribution of imperfection in

conductive constructal designs of arbitrary configurations". J Appl Phys. 112: 104905. Bibcode:2012JAP...112j4905E. doi:10.1063/1.4766443.

36. Pouzesh, M. R. Hajmohammadi and S. Poozesh, Investigations on the internal shape of constructal cavities intruding a heat generating body, Thermal Science, DOI: 10.2298/TSCI120427164P 2012.

37. M.R. Hajmohammadi, O. Joneydi Shariatzadeh, M. Moulod and S.S. Nourazar, Phi and Psi shaped conductive routes for improved cooling in a heat generating piece, International Journal of Thermal Sciences 2014; 77 66–74

38. Hajmohammadi, M. R.; Rahmani, M.; Campo, A.; Shariatzadeh, O. Joneydi (2014). "Optimal design of unequal heat flux elements for optimized heat transfer inside a rectangular duct". Energy. 68: 609–616. doi:10.1016/j.energy.2014.02.011.

39. Hajmohammadi, M. R.; Nourazar, S. S.; Campo, A.; Poozesh, S. (2013). "Optimal discrete distribution of heat flux elements for in-tube laminar forced convection". International Journal of Heat and Fluid Flow. 40: 89–96. doi:10.1016/j.ijheatfluidflow.2013.01.010.

40. Hajmohammadi, M. R.; Salimpour, M. R.; Saber, M.; Campo, A. (2013). "Detailed analysis for the cooling performance enhancement of a heat source under a thick plate". Energy Conversion and Management. 76: 691–700. doi:10.1016/j.enconman.2013.08.016.

41. Hajmohammadi, M. R.; Poozesh, S.; Hosseini, R. (2012). "Radiation effect on constructal design analysis of a T-Y-shaped assembly of fins". Journal of Thermal Science and Technology. 7: 677–692. Bibcode:2012JJTST...7..677H. doi:10.1299/jtst.7.677.

42. M. R. Errera and C. A. Marin, A Comparison Between Random and Deterministic Dynamics of River Drainage Basins Formation, Engenharia Térmica (Thermal Engineering), Vol. 8, n. 01, 65-71 (2009).

43. Hart, R. A. (2011). "Experimental thermal-hydraulic evaluation of constructal microfluidic structures under fully constrained conditions". Int J Heat Mass Transfer. 54: 3661–3671. doi:10.1016/j.ijheatmasstransfer.2011.02.063.

44. Hart, R. A.; Ponkala, M. J. V. (2011). "Development and testing of a constructal microchannel flow system with dynamically controlled complexity". Int J Heat Mass Transfer. 54: 5470–5480. doi:10.1016/j.ijheatmasstransfer.2011.07.044.

45. Z. Fan, X. Zhou, L. Luo and W. Yuan, "Experimental investigation of the flow distribution of a 2-dimensional constructal distributor. Exp Therm Fluid Sci 33, 77−83 (2008).

46. Cho, K.-H.; Chang, W.-P.; Kim, M.-H. (2011). "A numerical and experimental study to evaluate performance of vascularized cooling plates". Int J Heat Mass Transfer. 32: 1186–1198. doi:10.1016/j.ijheatfluidflow.2011.09.006.

47. D. Haller, P. Woias and N. Kockmann, "Simulation and experimental investigation of pressure loss and heat transfer in microchannel networks containing bends and T-junctions. Int J Heat Mass Transfer 52, 2678−2689 (2009).

48. L. Chen, "Progress in study on constructal theory and its applications. Science

China, Technological Sciences 55 (3), 802-820 (2012).

49. Bejan, Adrian; Reis, Antonio H. (25 March 2005). "Thermodynamic optimization of global circulation and climate". International Journal of Energy Research. 29 (4): 303–316. doi:10.1002/er.1058.
50. Bejan, Adrian (2009). "The Golden Ratio Predicted: Vision, Cognition And Locomotion As A Single Design In Nature" (PDF). International Journal of Design & Nature and Ecodynamics. 4 (2): 97–104. doi:10.2495/D&NE-V4-N2-97-104.
51. Bejan, Adrian (2007). "Why University Rankings do not Change:Education as a Natural Hierarchical Flow Architecture". Int. Journal of Design & Nature and Ecodynamics. 2 (4): 319–327. doi:10.2495/D&N-V2-N4-319-327.
52. Bejan, Adrian (Summer 1996). "Street network theory of organization in nature". Journal of Advanced Transportation. 30 (2): 85–107. doi:10.1002/atr.5670300207.
53. Bejan, Adrian; Ledezma, Gustavo A. (15 June 1998). "Streets tree networks and urban growth: Optimal geometry for quickest access between a finite-size volume and one point". Physica A: Statistical Mechanics and its Applications. 255 (1-2): 211–217. Bibcode:1998PhyA..255..211B. doi:10.1016/S0378-4371(98)00085-5.
54. Bejan, A.; Lorente, S.; Yilbas, B.S.; Sahin, A.Z. (26 April 2013). "Why solidification has an S-shaped history". Nature Scientific Reports. 3: 1711. Bibcode:2013NatSR...3E1711B. doi:10.1038/srep01711.
55. Bejan, Adrian; Marden, James H. (January 15, 2006). "Unifying constructal theory for scale effects in running, swimming and flying". Journal of Experimental Biology. 209 (2): 238–248. doi:10.1242/jeb.01974.
56. Charles, Jordan D.; Bejan, A. (August 1, 2009). "The evolution of speed, size and shape in modern athletics". Journal of Experimental Biology. 212 (15): 2419–2425. doi:10.1242/jeb.031161.
57. Bejan, Adrian (February 2001). "The tree of convective heat streams: its thermal insulation function and the predicted 3/4-power relation between body heat loss and body size". International Journal of Heat and Mass Transfer. 44 (4): 699–704. doi:10.1016/S0017-9310(00)00138-1.
58. Bejan, Adrian (June 1997). "Theory of Organization in Nature: Pulsating Physiological Processes". International Journal of Heat and Mass Transfer. 40 (9): 2097–2104. doi:10.1016/S0017-9310(96)00291-8.
59. Reis, A. H.; Miguel, A. F.; Aydin, M. (16 April 2004). "Constructal theory of flow architecture of the lungs". Medical Physics. 31 (5): 1135–1140. Bibcode:2004MedPh..31.1135R. doi:10.1118/1.1705443.
60. Bejan, A. (24 Aug 2012). "Why the bigger live longer and travel farther: animals, vehicles, rivers and the winds". Nature Scientific Reports. 2: 594. Bibcode:2012NatSR...2E.594B. doi:10.1038/srep00594. PMC 3426796Freely accessible. PMID 22924107.
61. Bejan, A.; Charles, J. D.; Lorente, S. (22 July 2014). "The evolution of airplanes". Journal of Applied Physics. 116: 044901. Bibcode:2014JAP...116d4901B. doi:10.1063/1.4886855.

62. Kleidon, A (2009). "Nonequilibrium thermodynamics and maximum entropy production in the Earth system: Applications and implications". Naturwissenschaften. 96: 653–677. Bibcode:2009NW....tmp...18K. doi:10.1007/s00114-009-0509-x.

63. Kleidon, A; Malhi, Y; Cox, PM (2010). "Maximum entropy production in environmental and ecological systems". Philos Trans R Soc Lond B Biol Sci. 365: 1297–1302. doi:10.1098/rstb.2010.0018.

64. Dewar, Roderick (7 January 2003). "Information theory explanation of the fluctuation theorem, maximum entropy production and self-organized criticality in non-equilibrium stationary states". Journal of Physics A: Mathematical and General. 36 (3): 631–641. doi:10.1088/0305-4470/36/3/303.

65. Dewar, Roderick C (10 May 2005). "Maximum entropy production and the fluctuation theorem". Journal of Physics A: Mathematical and General. 38 (21): L371-L381. doi:10.1088/0305-4470/38/21/l01.

66. Bejan, A. (2010). Design in nature, thermodynamics, and the constructal law. Comment on "Life, hierarchy, and the thermodynamic machinery of planet Earth" by A. Kelidon, Physics of Life Reviews Vol. 7, No. 4: 467-470.

67. Rocha, L. A. O. (2011). "Constructal law: from the law of physics to applications and conferences". Phys Life Rev. 8: 245–246. doi:10.1016/j.plrev.2011.07.005.

68. Bejan, A; Lorente, S (2011). "The constructal law and the evolution of design in nature". Physics of Life Reviews. 8 (3): 209–240. doi:10.1016/j.plrev.2011.05.010.

69. Bejan and S. Lorente, Constructal law of design and evolution: Physics, biology, technology, and society, Journal of Applied Physics 113, 151301 (2013); doi:10.1063/1.4798429

70. Bejan, A; Lorente, S (2004). "The Constructal Law and the Thermodynamics of Flow Systems with Configuration". International Journal of Heat and Mass Transfer. 47: 3203–3214. doi:10.1016/j.ijheatmasstransfer.2004.02.007.

71. Bejan, Adrian; Zane, Peder (January 24, 2012). Design in Nature: How the Constructal Law Governs Evolution in Biology, Physics, Technology, and Social Organization. New York: Doubleday. p. 11. ISBN 978-0-307-744340.

72. Bejan, Adrian (May 24, 2016). The Physics of Life: The Evolution of Everything. St. Martin's Press. p. 272. ISBN 1250078822.

73. Reis, Antonio Heitor. "Use and validity of principles of extremum of entropy production in the study of complex systems". Annals of Physics. 346: 22–27. doi:10.1016/j.aop.2014.03.013.

CHAPTER 6

Ecological Forecasting

Chapter Contents

Ecological Forecasting employs observations and models to predict the impacts of environmental change on ecosystems. In doing so, it applies information from the physical, biological, and social sciences and promotes a scientific synthesis across the domains of physics, geology, chemistry, biology, and psychology. The goal is reliable forecasts that allow decision makers access to science-based tools in order to project changes in living systems. These forecasts should incorporate knowledge of uncertainties and estimates of error and allow those making decisions to compare the outcomes of alternative policies.

INTRODUCTION

Ecological forecasting uses knowledge of physics, ecology and physiology to predict how ecosystems will change in the future in response to environmental factors such as climate change. The ultimate goal of the approach is to provide people such as resource managers and designers of marine reserves with information that they can then use to respond, in advance, to future changes, a form of adaptation to global warming.

One of the most important environmental factors for organisms today is global warming. Most physiological processes are affected by temperature, and so even small changes in weather and climate can lead to large changes in the growth, reproduction and survival of animals and plants. The scientific consensus is that the increase in atmospheric greenhouse gases due to human activity caused most of the warming observed since the start of the industrial era. These changes are in turn affecting human and natural ecosystems.

One major challenge is to predict where, when and with what magnitude changes are likely to occur so that we can mitigate or at least prepare for them. Ecological forecasting applies existing knowledge of how animals and plants interact with their physical environment to ask how changes in environmental factors might result in changes to the ecosystems as a whole.

APPROACHES

- Palaeobiology modeling: uses fossil and phylogenetic evidence of biodiversity in the past to project the trajectory of biodiversity in the future. Simple plots can be constructed and then adjusted based on the varying quality of the fossil record.
- Climate envelope modeling: relies on statistical correlations between existing species distributions and environmental variables to define a species' tolerance. *Envelopes* of tolerance are then drawn around existing ranges. By predicting future levels of factors such as temperature, rainfall, and salinity, new range boundaries are then predicted. These methods are good for examining large numbers of species, but are likely not a good means of predicting effects at fine scales.
- Niche level modeling: is a newer method which links physiological information about a species to models of animal and plant body temperature. In contrast to "climate envelope" approaches, environmental variables are predicted at the level of the niche and are therefore much more exact. However, the approach is also usually more time consuming.

FORECASTING EXAMPLES

Biodiversity

Using fossil evidence, studies have shown that vertebrate biodiversity has grown exponentially through Earth's history and that biodiversity is entwined with the diversity of Earth's habitats.

"Animals have not yet invaded 2/3 of Earth's habitats, and it could be that without human influence biodiversity will continue to increase in an exponential fashion."

—Sahney et al.

Temperature

Forecasts of temperature, shown in the diagram at the right as colored dots, along the North Island of New Zealand in the austral summer of 2007. As per the temperature scale shown at the bottom, intertidal temperatures were forecast to exceed 30 °C at some locations on February 19; surveys later showed that these sites corresponded to large die-offs in burrowing sea urchins.

ECOLOGICAL HUMANITIES

The ecological humanities (also environmental humanities) is an interdisciplinary area of research, drawing on the many environmental sub-disciplines that have emerged in the humanities over the past several decades (in particular environmental literature, environmental philosophy, environmental history and environmental anthropology). The ecological humanities aim to help bridge traditional divides between the sciences and the humanities, and between Western, Eastern and Indigenous ways of knowing the natural world and the place of humans in it (Rose 2004).

The ecological humanities are characterised by a *connectivity ontology* and a commitment to two fundamental axioms relating to the need to submit to ecological laws and to see humanity as part of a larger living system.

Connectivity Ontology

One of the fundamental ontological presuppositions of ecological humanities is that the organic world and its inorganic parts are seen as a single system whereby each part is linked to each other part. This world view in turn shares an intimate connection with Lotka›s physiological philosophy and the associated concept of the «World Engine». When we see everything as connected, then the traditional questions of the humanities concerning economic and political justice become *enlarged,* into a consideration of how justice is connected with our transformation of our environment and ecosystems. The consequence of such a connectivity ontology is, as proponents of the ecological humanities argue, that we begin to seek out a more inclusive concept of justice that includes non-humans within the domain of those to whom rights are owing. This broadened conception of justice involves "enlarged" or "ecological thinking", which presupposes the enhancement of knowledge sharing within fields of plural and diverse 'knowledges'. This kind of knowledge sharing is called transdisciplinarity. It has links with the political philosophy of Hannah Arendt and the works of Italo Calvino. As Calvino put it, «enlarge[s] the sphere of what we can imagine». It also has connections with Leibniz›s Enlightenment project where, the sciences are simultaneously abridged while also being enlarged.

The situation is complicated however by the recognition of the fact that connections are both *non-linear* and *linear*. The ecological humanities therefore, require both linear and non-linear modes of language through which reasoning about justice can be done. Thus there is a motivation to find linguistic modes which can adequately express both linear and non-linear connectivities.

Axioms of Ecological Humanities

There are three axioms of ecological humanities :

- The axiom of submission to ecosystem laws;
- The axiom of ecological kinship, which situates humanity as participant in a larger living system; and
- The axiom of the social construction of ecosystems and ecological unity, which states that ecosystems and nature may be merely convenient conceptual entities (Marshall, 2002).

Putting the first and second axioms another way, the connections between and among living things are the basis for how ecosystems are understood to work, and thus constitute laws of existence and guidelines for behaviour (Rose 2004)

The first of these axioms has a tradition in social sciences (*see* Marx, 1968: 3). From the second axiom the notions of "ecological embodiment/ embededness" and "habitat" have emerged from Political Theory with a fundamental connectivity to rights, democracy and ecologism (Eckersley 1996: 222, 225; Eckersley 1998).

The third axiom comes from the strong ‹self-reflective› tradition of all ‹humanities› scholarship and it encourages ecological humanities to investigate its own theoretical basis (and without which, ecological humanities is just ‹ecology›).

Contemporary Ideas

Political economic ecology

Some theorists have suggested that the inclusion of non-humans in the consideration of justice links ecocentric philosophy with political economics. This is because the theorising of justice is a central activity of political economic philosophy. If in accordance with the axioms of ecological humanities, theories of justice are enlarged to include ecological values then the necessary result is the synthesis of the concerns of ecology with that of political economy: i.e. Political Economic Ecology.

Energy Systems Language

The question of what language can best depict the linear and non-linear causal connections of ecological systems appears to have been taken up by the school of ecology known as systems ecology. To depict the linear and non-linear internal relatedness of ecosystems where the laws of thermodynamics hold significant consequences (Hannon et al. 1991: 80), Systems Ecologist H.T. Odum (1994) predicated the Energy Systems Language on the principles of ecological energetics. In ecological energetics, just as in ecological humanities, the causal bond between connections is considered an ontic category (see Patten et al. 1976: 460). Moreover, as a result of simulating ecological systems with the energy systems language H.T.Odum make the controversial suggestion that embodied energy could be understood as value, which in itself is a step into the field of Political Economic Ecology noted above.

REFERENCES

1. Clark et al. 2001
2. «Joint science academies› statement: The science of climate change». Royal Society. 2001-05-17. Archived from the original on October 1, 2007. Retrieved 2007-04-01. The work of the Intergovernmental Panel on Climate Change (IPCC) represents the consensus of the international scientific community on climate change science
3. «Rising to the climate challenge». Nature. 449 *(7164): 755. 2007-10-18. doi*:10.1038/449755a. PMID 17943093. Retrieved 2007-11-06.
4. CCSP 2008
5. Kearney 2006
6. Gilman et al. 2006
7. Wethey and Woodin 2008
8. ^ Jump up to:[a] [b] Sahney, S.; Benton, M.J. & Ferry, P.A. (2010). «Links between global taxonomic diversity, ecological diversity and the expansion of vertebrates on land» (PDF). Biology Letters. 6 *(4): 544–547. doi:10.1098/rsbl.2009.1024. PMC 2936204. PMID 20106856.*
9. Pearson and Dawson 2003
10. Kearney et al. 2008
11. Helmuth et al. 2006
12. Italo Calvino, On Fourier, III: A Utopia of Fine Dust, The Literature Machine, Picador, London.
13. R. Eckersley (1996) 'Greening Liberal Democracy', in Doherty, B. and de Geus, M. ed. Democracy & Green Political Thought: Sustainability, Rights and Citizenship, Routledge, London, pp. 212–236.
14. R. Eckersley (1998) 'The Death of Nature and the Birth of Ecological Humanities', Organization and Environment, Vol 11, No. 2, pp. 183–185.
15. R. Eckersley (2001) ‹Symposium Green Thinking – from Australia›, Environmental Politics, Vol.10, No.4, pp. 85–102.
16. J.B. Foster and P.Burkett (2004) 'Ecological Economics And Classical Marxism', Organization & Environment, Vol. 17, No.1, pp. 32–60.
17. B. Hannon, R.Costanza and R.Ulanowicz (1991) 'A General Accounting Framework for Ecological Systems: A Functional Taxonomy for Connectivist Ecology', Theoretical Population Biology, Vol. 40, 78-104.
18. Marshall (2002) The Unity of Nature: Wholeness and Disintegration in Ecology and Science. London: Imperial College Press.
19. J. Martinez-Alier (1987) Ecological Economics, Basil Blackwell.
20. K. Marx (1968), in Karl Marx: 1818/1968, a collection of essays, Inter Nationes, Bad Godesberg.

21. H.T. Odum (1994) Ecological and General Systems: An Introduction to Systems Ecology, Colorado University Press, Boulder, Colorado.
22. B.C. Patten, R.W.Bosserman, J.T.Finn and W.B.Cale (1976) 'Propagation of Cause in Ecosystems', in Patten, B.C. ed. Systems Analysis and Systems Simulation in Ecology, Academic Press inc. New York.
23. S. Podolinsky (2004) 'Socialism And The Unity Of Physical Forces', Organization & Environment, Vol. 17, No. 1, pp. 61–75.
24. D. Rose and L. Robin (2004) ‹The Ecological Humanities in Action: An Invitation›, Australian Humanities Review, 31-2
25. D.R. Weiner (2000) Models of Nature: Ecology, Conservation and Cultural Revolution in Soviet Russia, University of Pittsburgh Pre

CHAPTER 7

Ecological Stoichiometry

Chapter Contents

If you thought you left balancing chemical reactions (stoichiometry) behind in freshman chemistry class, think again. Biological Stoichiometry is the study of the balance of energy and multiple chemical elements in living systems. This approach uses the same concepts and principles as chemical stoichiometry, but applies them to many phenomena at all levels of biology, from genes and molecules to whole organisms and even to ecosystems and the biosphere. Biological Stoichiometry is not a new concept; it emerged from Ecological Stoichiometry, which itself has deep intellectual roots in the work of the famous scientists Justus Leibig, AJ Lotka, and AC Redfield. While the approach is general and can be applied to any set of elements, most work to date has focused on the major elements carbon (C), nitrogen (N), and phosphorus (P).

INTRODUCTION

Ecological stoichiometry considers how the balance of energy and elements affects and is affected by organisms and their interactions in ecosystems. Ecological stoichiometry has a long history in ecology with early references to the constraints of mass balance made by Liebig, Lotka, and Redfield, and has recently gained momentum by explicitly linking the elemental physiology of organisms to their food web interactions and ecosystem function.

Most work in ecological stoichiometry focuses on the interface between a consumer and its food. This interface, whether it is between plants and their resources or large herbivores and grasses, is often characterized by dramatic differences in the elemental composition of each participant. Consider termites which have a body C:N of about 5 but consume wood with a C:N ratio of 300-1000. Ecological stoichiometry primarily asks:

- why do elemental imbalances arise in nature?
- how is consumer physiology and life-history affected by elemental imbalances? and
- what are the subsequent effects on ecosystem processes?

Elemental imbalances are a mismatch between the elemental demands of a consumer and that present in its resources. Elemental imbalances arise between grazers and their food whose foods vary considerably in their elemental composition more often than in animals who have less elemental flexibility. For example, carbon to phosphorus ratios in the suspended organic matter in lakes (i.e., algae, bacteria, and detritus) can vary between 100 and 1000 whereas C:P ratios of *Daphnia*, a crustacean zooplankton, remain nearly constant at 80:1. There are a number of physiological and evolutionary explanations for these differences in elemental composition that are related to the types of needed resources, their relative availability in time and space, and how they are acquired.

The degree to which organisms maintain a constant chemical composition in the face of variations in their environment, particularly in the chemical composition and availability of their resources, is referred to as "stoichiometric homeostasis". Like the general biological notion of homeostasis, elemental homeostasis refers to the maintenance of elemental composition within some biologically ordered range. Photoautotrophic organisms, such as algae and vascular plants, can exhibit a very wide range of physiological plasticity in elemental composition and thus have relatively weak stoichiometric homeostasis. In contrast, other organisms, multicellular animals for example, have close to strict homeostasis and they can be thought of as having distinct chemical composition.

Ecological stoichiometry seeks to discover how the chemical content of organisms shapes their ecology. Ecological stoichiometry has been applied to studies of nutrient recycling, resource competition, animal growth, and nutrient limitation patterns in whole ecosystems. The Redfield ratio of the world's oceans is one very famous application of stoichiometric principles to ecology. Ecological stoichiometry also considers phenomena at the sub-cellular level, such as the P-content of a ribosome, as well as phenomena at the whole biosphere level, such as the oxygen content of Earth's atmosphere.

ECOPATH

Ecopath with Ecosim (EwE) is a free ecosystem modelling software suite, initially started at NOAA by Jeffrey Polovina, but has since primarily been developed at the UBC Fisheries Centre of the University of British Columbia. In 2007, it was named as one of the ten biggest scientific breakthroughs in NOAA's 200-year history. The NOAA citation states that Ecopath "revolutionized scientists' ability worldwide to understand complex marine ecosystems".Behind this lie more than two decades of development work in association with Villy Christensen, Carl Walters, Daniel Pauly, and other fisheries scientists,followed with the provision of user support, training and co-development collaborations. Per January 2012 there are 6000+ registered users in 150+ countries.

Components

EwE has three main components:

- Ecopath – a static, mass-balanced snapshot of the system
- Ecosim – a time dynamic simulation module for policy exploration
- Ecospace – a spatial and temporal dynamic module primarily designed for exploring impact and placement of protected areas.

Capabilities

The Ecopath software package can be used to:

- address ecological questions
- evaluate ecosystem effects of fishing
- explore management policy options
- analyze impact and placement of marine protected areas
- predict movement and accumulation of contaminants and tracers (Ecotracer)
- model effect of environmental changes,
- facilitate end-to-end ecosystem model construction

The desktop version of Ecopath with Ecosim runs only on Windows and requires Microsoft Access database drivers version 2007 or newer. The computational core of Ecopath with Ecosim can be executed on other operating systems such as Unix or Linux using the Mono common language runtime.

Development Ecopath version 6 received support from the Lenfest Ocean Program and the Pew Charitable trusts. In 2011 the Ecopath Research and Development Consortium was founded to share the responsibility of maintaining and further developing the approach with institutions around the world.

REFERENCES

1. NOAA, 2007. ECOPATH Modeling: Precursor to an Ecosystem Approach to Fisheries Management [WWW Document]. URL http://celebrating200years.noaa.gov/breakthroughs/ecopath/welcome.html (accessed 8.26.12).
2. NOAA, 2007. ECOPATH Modeling: Precursor to an Ecosystem Approach to Fisheries Management [WWW Document]. URL http://celebrating200years.noaa.gov/breakthroughs/ecopath/welcome.html (accessed 8.26.12).
3. Smikle, S.G., V. Christensen and K.A. Aiken (2010). "A review of caribbean ecosystems and fishery resources using ECOPATH models". Revue Etudes Caribéennes, n°15, https://etudescaribeennes.revues.org/4529

4. Christensen, V. and Pauly, D., 1992. Ecopath II - a software for balancing steady-state ecosystem models and calculating network characteristics. Ecological Modelling, 61:169-185.

5. Walters, C., Christensen, V. and Pauly, D., 1997. Structuring dynamic models of exploited ecosystems from trophic mass-balance assessments. Rev Fish Biol Fish, 7:139-172

6. Walters, C., Pauly, D. and Christensen, V., 1999. Ecospace: Prediction of mesoscale spatial patterns in trophic relationships of exploited ecosystems, with emphasis on the impacts of marine protected areas. Ecosystems, 2:539-554.

7. Christensen, V. and Pauly, D. (Editors), 1993. Trophic Models of Aquatic Ecosystems. ICLARM Conference Proceedings 26, Manila, 390 p.

8. Walters, C.J., Christensen, V., Martell, S.J. and Kitchell, J.F., 2005. Possible ecosystem impacts of applying MSY policies from single-species assessment. ICES J Mar Sci, 62:558-568.

9. Christensen, V. and Walters, C.J., 2004. Trade-offs in ecosystem-scale optimization of fisheries management policies. Bull Mar Sci, 74:549-562

10. Walters, C.J. and Martell, S.J.D., 2004. Fisheries Ecology and Management. Princeton University Press, Princeton, 399 p.

11. Christensen, V. and Booth, S., 2006. Ecosystem modeling of dioxin distribution patterns in the marine environment. Chapter 6. In: J. Alder and D. Pauly (Editor), On the multiple uses of small pelagic fishes: from ecosystems to markets. UBC Fisheries Centre Research Reports 14(3). Fisheries Centre, University of British Columbia [ISSN 1198-6727], Vancouver

12. Guénette, S., Heymans, J.J., Christensen, V. and Trites, A.W., 2006. Ecosystem models show combined effects of fishing, predation, competition, and ocean productivity on Steller sea lions (Eumetopias jubatus) in Alaska. Can J Fish Aquat Sci, 63:2495-2517.

13. Christensen, V., Coll, M., Steenbeek, J., Buszowski, J., Chagaris, D., and Walters, C.J., 2014. Representing Variable Habitat Quality in a Spatial Food Web Model. Ecosystems 17, 1397–1412.

14. Steenbeek, J., Coll, M., Gurney, L., Mélin, F., Hoepffner, N., Buszowski, J., and Christensen, V., 2013. Bridging the gap between ecosystem modelling tools using geographic information systems: driving a food-web model with spatial-temporal primary production data. Ecological Modelling 263, 139–151.

15. *Steenbeek, Jeroen; Buszowski, Joe; Christensen, Villy; Akoglu, Ekin; Aydin, Kerim; Ellis, Nick; Felinto, Dalai; Guitton, Jerome; Lucey, Sean; Kearney, Kelly; Mackinson, Steven; Pan, Mike; Platts, Mark; Walters, Carl (January 2016). "Ecopath with Ecosim as a model-building toolbox: Source code capabilities, extensions, and variations". Ecological Modelling.* ***319****: 178–189. doi:10.1016/j.ecolmodel.2015.06.031.*

16. Christensen, V. and Lai, S., 2007. Ecopath with Ecosim 6: the sequel. The Sea

Around Us Project Newsletter, 43:1-4 (September–October).

17. Smikle, S.G., V. Christensen and K.A. Aiken (2010). «A review of caribbean ecosystems and fishery resources using ECOPATH models». Revue Etudes Caribéennes, n°15, https://etudescaribeennes.revues.org/4529

18. Christensen, V. and Pauly, D., 1992. Ecopath II - a software for balancing steady-state ecosystem models and calculating network characteristics. Ecological Modelling, 61:169-185.

19. Walters, C., Christensen, V. and Pauly, D., 1997. Structuring dynamic models of exploited ecosystems from trophic mass-balance assessments. Rev Fish Biol Fish, 7:139-172

20. Walters, C., Pauly, D. and Christensen, V., 1999. Ecospace: Prediction of mesoscale spatial patterns in trophic relationships of exploited ecosystems, with emphasis on the impacts of marine protected areas. Ecosystems, 2:539-554.

21. Christensen, V. and Pauly, D. (Editors), 1993. Trophic Models of Aquatic Ecosystems. ICLARM Conference Proceedings 26, Manila, 390 p.

22. Walters, C.J., Christensen, V., Martell, S.J. and Kitchell, J.F., 2005. Possible ecosystem impacts of applying MSY policies from single-species assessment. ICES J Mar Sci, 62:558-568.

23. Christensen, V. and Walters, C.J., 2004. Trade-offs in ecosystem-scale optimization of fisheries management policies. Bull Mar Sci, 74:549-562

24. Walters, C.J. and Martell, S.J.D., 2004. Fisheries Ecology and Management. Princeton University Press, Princeton, 399 p.

25. Christensen, V. and Booth, S., 2006. Ecosystem modeling of dioxin distribution patterns in the marine environment. Chapter 6. In: J. Alder and D. Pauly (Editor), On the multiple uses of small pelagic fishes: from ecosystems to markets. UBC Fisheries Centre Research Reports 14(3). Fisheries Centre, University of British Columbia [ISSN 1198-6727], Vancouver

26. Guénette, S., Heymans, J.J., Christensen, V. and Trites, A.W., 2006. Ecosystem models show combined effects of fishing, predation, competition, and ocean productivity on Steller sea lions (Eumetopias jubatus) in Alaska. Can J Fish Aquat Sci, 63:2495-2517.

27. Christensen, V., Coll, M., Steenbeek, J., Buszowski, J., Chagaris, D., and Walters, C.J., 2014. Representing Variable Habitat Quality in a Spatial Food Web Model. Ecosystems 17, 1397–1412.

28. Steenbeek, J., Coll, M., Gurney, L., Mélin, F., Hoepffner, N., Buszowski, J., and Christensen, V., 2013. Bridging the gap between ecosystem modelling tools using geographic information systems: driving a food-web model with spatial-temporal primary production data. Ecological Modelling 263, 139–151.

29. *Steenbeek, Jeroen; Buszowski, Joe; Christensen, Villy; Akoglu, Ekin; Aydin, Kerim; Ellis, Nick; Felinto, Dalai; Guitton, Jerome; Lucey, Sean; Kearney, Kelly; Mackinson, Steven; Pan, Mike; Platts, Mark; Walters, Carl (January 2016). «Ecopath with Ecosim as a model-building toolbox: Source code capabilities,*

extensions, and variations». Ecological Modelling. ***319****: 178–189. doi:10.1016/j. ecolmodel.2015.06.031.*

30. Christensen, V. and Lai, S., 2007. Ecopath with Ecosim 6: the sequel. The Sea Around Us Project Newsletter, 43:1-4 (September–October).
31. R. W. Sterner and J. J. Elser (2002) Ecological Stoichiometry: The Biology of Elements from Molecules to the Biosphere. Princeton University Press. pp.584. ISBN 0691074917
32. Olff, H; Alonso, D; Berg, MP; Eriksson, BK; Loreau, M; Piersma, T; Rooney, N (2009). "Parallel ecological networks in ecosystems". Phil. Trans. R. Soc. B. 364 (1755–1779): 1502–4. doi:10.1098/rstb.2008.0222.
33. Martinson, H. M., K. Schneider, J. Gilbert, J. Hines, P. A. Hambäck, W. F. Fagan. 2008. Detritivory: Stoichiometry of a neglected trophic level. Ecological Research 23: 487-491 DOI: 10.1007/s11284-008-0471-7

CHAPTER 8

Fisheries Fish Ecology and Climate Change

Chapter Contents

- Introduction
- History
- Political Objectives
- International Objectives
- Management Mechanisms
- Catch Quotas
- Fisheries Law
- Climate Change
- Population Dynamics
- Ecosystem Based Fisheries
- Elderly Maternal Fish
- Data Quality
- Ecopath
- Human Factors
- Performance
- Endolith

Fish and fisheries are an integral part of most societies and make important contributions to economic and social health and well-being in many countries and areas. It has been estimated that approximately 12.5 million people are employed in fishery-related activities, and in recent years global production from capture fisheries has tended to vary between approximately 85 and 90 million tonnes. The products from these fisheries are used in a wide variety of ways, ranging from subsistence use to international trade as highly sought-after and highly-valued items. The value of fish traded internationally is approximately US$40 billion per year.

INTRODUCTION

Fisheries management draws on fisheries science in order to find ways to protect fishery resources so sustainable exploitation is possible. Modern fisheries management is often referred to as a governmental system of appropriate management rules based on defined objectives and a mix of management means to implement the rules, which are put in place by a system of monitoring control and surveillance. According to the Food and Agriculture Organization of the United Nations (FAO), there are no clear and generally accepted definitions of fisheries management. However, the working definition used by the FAO and much cited elsewhere is:

The integrated process of information gathering, analysis, planning, consultation, decision-making, allocation of resources and formulation and implementation, with enforcement as necessary, of regulations or rules which govern fisheries activities in order to ensure the continued productivity of the resources and the accomplishment of other fisheries objectives.

HISTORY

Fisheries have been explicitly managed in some places for hundreds of years. More than 80 percent of the worlds commercial exploitation of fish and shellfish are harvest from natural occurring populations in the oceans and freshwater areas. For example, the Māori people, New Zealand residents for about 700 years, had prohibitions against taking more than what could be eaten and about giving back the first fish caught as an offering to sea god Tangaroa. Starting in the 18th century attempts were made to regulate fishing in the North Norwegian fishery. This resulted in the enactment of a law in 1816 on the Lofoten fishery, which established in some measure what has come to be known as territorial use rights.

The fishing banks were divided into areas belonging to the nearest fishing base on land and further subdivided into fields where the boats were allowed to fish. The allocation of the fishing fields was in the hands of local governing committees, usually headed by the owner of the onshore facilities which the fishermen had to rent for accommodation and for drying the fish.

Governmental resource protection-based fisheries management is a relatively new idea, first developed for North European fisheries after the first Overfishing Conference held in London in 1936. In 1957 British fisheries researchers Ray Beverton and Sidney Holt published a seminal work on North Sea commercial fisheries dynamics. In the 1960s the work became the theoretical platform for North European management schemes.

After some years away from the field of fisheries management, Beverton criticized his earlier work in a paper given at the first World Fisheries Congress in Athens in 1992. The Dynamics of Exploited Fish Populations expressed his concerns, including the way his and Sidney Holts work had been misinterpreted and misused by fishery biologists and managers during the previous 30 years. Nevertheless, the institutional foundation for modern fishery management had been laid.

In 1996, the Marine Stewardship Council was founded to set standards for sustainable fishing. In 2010, the Aquaculture Stewardship Council was created to do the same for aquaculture.

A report by Prince Charles International Sustainability Unit, the New York-based Environmental Defense Fund and 50in10 published in July 2014 estimated global fisheries were adding $270 billion a year to global GDP, but by full implementation of sustainable fishing, that figure could rise by an extra amount of as much as $50 billion.

POLITICAL OBJECTIVES

According to the FAO, fisheries management should be based explicitly on political objectives, ideally with transparent priorities. Typical political objectives when exploiting a fish resource are to:

- maximize sustainable biomass yield
- maximize sustainable economic yield
- secure and increase employment
- secure protein production and food supplies
- increase export income

Such political goals can also be a weak part of fisheries management, since the objectives can conflict with each other.

INTERNATIONAL OBJECTIVES

Fisheries objectives need to be expressed in concrete management rules. In most countries fisheries management rules should be based on the internationally agreed, though non-binding, Code of Conduct for Responsible Fisheries, agreed at a meeting of the U.N.'s Food and Agriculture Organization FAO session in 1995. The precautionary approach it prescribes is typically implemented in concrete management rules as minimum spawning biomass, maximum fishing mortality rates, etc. In 2005 the UBC Fisheries Centre at the University of British Columbia comprehensively reviewed the performance of the world's major fishing nations against the Code.

International agreements are required in order to regulate fisheries in international waters. The desire for agreement on this and other maritime issues led to three conferences on the Law of the Sea, and ultimately to the treaty known as the United Nations Convention on the Law of the Sea (UNCLOS). Concepts such as exclusive economic zones (EEZ, extending 200 nautical miles (370 km) from a nations coasts) allocate certain sovereign rights and responsibilities for resource management to individual countries.

Other situations need additional intergovernmental coordination. For example, in the Mediterranean Sea and other relatively narrow bodies of water, EEZ of 200 nautical miles (370 km) are irrelevant. International waters beyond 12-nautical-mile (22 km) from shore require explicit agreements.

Straddling fish stocks, which migrate through more than one EEZ also present challenges. Here sovereign responsibility must be agreed with neighbouring coastal states and fishing entities. Usually this is done through the medium of a regional organisation set up for the purpose of coordinating the management of that stock.

UNCLOS does not prescribe precisely how fisheries confined only to international waters should be managed. Several new fisheries (such as high seas bottom trawling fisheries) are not (yet) subject to international agreement across their entire range. In November 2004 the UN General Assembly issued a resolution on Fisheries that prepared for further development of international fisheries management law.

MANAGEMENT MECHANISMS

Many countries have set up Ministries/Government Departments, named "Ministry of Fisheries" or similar, controlling aspects of fisheries within their exclusive economic zones. Four categories of management means have been devised, regulating either input/ investment, or output, and operating either directly or indirectly:

	Inputs	Outputs
Indirect	Vessel licensing	Catching techniques
Direct	Limited entry	Catch quota and technical regulation

Technical means may include:

- prohibiting devices such as bows and arrows, and spears, or firearms
- prohibiting nets
- setting minimum mesh sizes
- limiting the average potential catch of a vessel in the fleet (vessel and crew size, gear, electronic gear and other physical "inputs".
- prohibiting bait
- snagging
- limits on fish traps
- limiting the number of poles or lines per fisherman
- restricting the number of simultaneous fishing vessels
- limiting a vessel's average operational intensity per unit time at sea
- limiting average time at sea

CATCH QUOTAS

Systems that use *individual transferable quotas* (ITQ), also called individual fishing quota limit the total catch and allocate shares of that quota among the fishers who work that fishery. Fishers can buy/sell/trade shares as they choose.

A large scale study in 2008 provided strong evidence that ITQs can help to prevent fishery collapse and even restore fisheries that appear to be in decline. Other studies have shown negative socio-economic consequences of ITQs, especially on small-sclale fisheries. These consequences include concentration of quota in that hands of few fishers; increased number of inactive fishers leasing their quotas to others (a phenomenon known as armchair fishermen); and detrimental effects on coastal communities.

Precautionary principle

The *Fishery Manager's Guidebook* issued in 2009 by the FAO of the United Nations, advises that the precautionary approach or principle should be applied when "ecosystem resilience and human impact (including reversibility) are difficult to forecast and hard to distinguish from natural changes." The precautionary principle suggests that when an action risks harm, it should not be proceeded with until it can be scientifically proven to be safe. Historically fishery managers have applied this principle the other way round; fishing activities have not been curtailed until it has been proven that they have already damaged existing ecosystems. In a paper published in 2007, Shertzer and Prager suggested that there can be significant benefits to stock biomass and fishery yield if management is stricter and more prompt.

FISHERIES LAW

Fisheries law is an emerging and specialized area of law which includes the study and analysis of different fisheries management approaches, including seafood safety regulations and aquaculture regulations. Despite its importance, this area is rarely taught at law schools around the world, which leaves a vacuum of advocacy and research.

CLIMATE CHANGE

In the past, changing climate has affected inland and offshore fisheries and such changes are likely to continue. From a fisheries perspective, the specific driving factors of climate change include rising water temperature, alterations in the hydrologic cycle, changes in nutrient fluxes, and relocation of spawning and nursery habitat. Further, changes in such factors would affect resources at all levels of biological organization, including the genetic, organism, population, and ecosystem levels.

POPULATION DYNAMICS

Population dynamics describes the growth and decline of a given fishery stock over time, as controlled by birth, death and migration. It is the basis for understanding changing fishery patterns and issues such as habitat destruction, predation and optimal harvesting rates. The population dynamics of fisheries has been traditionally used by fisheries scientists to determine sustainable yields.

The basic accounting relation for population dynamics is the BIDE model:

$$N_1 = N_0 + B - D + I - E$$

where N_1 is the number of individuals at time 1, N_0 is the number of individuals at time 0, B is the number of individuals born, D the number that died, I the number that immigrated, and E the number that emigrated between time 0 and time 1. While immigration and emigration can be present in wild fisheries, they are usually not measured.

Care is needed when applying population dynamics to real world fisheries. In the past, over-simplistic modelling, such as ignoring the size, age and reproductive status of the fish, focusing solely on a single species, ignoring bycatch and physical damage to the ecosystem, has accelerated the collapse of key stocks.

ECOSYSTEM BASED FISHERIES

According to marine ecologist Chris Frid, the fishing industry points to pollution and global warming as the causes of unprecedentedly low fish stocks in recent years, writing, "Everybody would like to see the rebuilding of fish stocks and this can only be achieved if we understand all of the influences, human and natural, on fish dynamics." Overfishing has also had an effect. Frid adds, "Fish communities can be altered in a number of ways, for example they can decrease if particular sized individuals of a species are targeted, as this affects predator and prey dynamics. Fishing, however, is not the sole perpetrator of changes to marine life - pollution is another example [...] No one factor operates in isolation and components of the ecosystem respond differently to each individual factor."

In contrast to the traditional approach of focusing on a single species, the ecosystem-based approach is organized in terms of ecosystem services. Ecosystem-based fishery concepts have been implemented in some regions.In 2007 a group of scientists offered the following *ten commandments*

"

- Keep a perspective that is holistic, risk-adverse and adaptive.
- Maintain an "old growth" structure in fish populations, since big, old and fat female fish have been shown to be the best spawners, but are also susceptible to overfishing.
- Characterize and maintain the natural spatial structure of fish stocks, so that management boundaries match natural boundaries in the sea.
- Monitor and maintain seafloor habitats to make sure fish have food and shelter.
- Maintain resilient ecosystems that are able to withstand occasional shocks.
- Identify and maintain critical food-web connections, including predators and forage species.
- Adapt to ecosystem changes through time, both short-term and on longer cycles of decades or centuries, including global climate change.
- Account for evolutionary changes caused by fishing, which tends to remove large, older fish.
- Include the actions of humans and their social and economic systems in all ecological equations.

"

Report to Congress (2009): The State of Science to Support an Ecosystem Approach to Regional Fishery ManagementNational Marine Fisheries Service, NOAA Technical Memorandum NMFS-F/SPO-96.

ELDERLY MATERNAL FISH

Old fat female rockfish are the best producers

Traditional management practices aim to reduce the number of old, slow-growing fish, leaving more room and resources for younger, faster-growing fish. Most marine fish produce huge numbers of eggs. The assumption was that younger spawners would produce plenty of viable larvae.

However, 2005 research on rockfish shows that large, elderly females are far more important than younger fish in maintaining productive fisheries. The larvae produced by these older maternal fish grow faster, survive starvation better, and are much more likely to survive than the offspring of younger fish. Failure to account for the role of older fish may help explain recent collapses of some major US West Coast fisheries. Recovery of some stocks is expected to take decades. One way to prevent such collapses is to establish marine reserves, where fishing is not allowed and fish populations age naturally.

DATA QUALITY

According to fisheries scientist Milo Adkison, the primary limitation in fisheries management decisions is the absence of quality data. Fisheries management decisions are often based on population models, but the models need quality data to be effective. He asserts that scientists and fishery managers would be better served with simpler models and improved data.

The most reliable source for summary statistics is the FAO Fisheries Department.

ECOPATH

Ecopath, with Ecosim (EwE), is an ecosystem modelling software suite. It was initially a NOAA initiative led by Jeffrey Polovina, later primarily developed at the UBC Fisheries Centre of the University of British Columbia. In 2007, it was named as one of the ten biggest scientific breakthroughs in NOAA's 200-year history. The citation states that Ecopath revolutionized scientists' ability worldwide to understand complex marine ecosystems. Behind this lies two decades of development work by Villy Christensen, Carl Walters, Daniel Pauly, and other fisheries scientists. As of 2010 there are 6000 registered users in

155 countries. Ecopath is widely used in fisheries management as a tool for modelling and visualising the complex relationships that exist in real world marine ecosystems.

HUMAN FACTORS

Managing fisheries is about managing people and businesses, and not about managing fish. Fish populations are managed by regulating the actions of people. If fisheries management is to be successful, then associated human factors, such as the reactions of fishermen, are of key importance, and need to be understood.

Management regulations must also consider the implications for stakeholders. Commercial fishermen rely on catches to provide for their families just as farmers rely on crops. Commercial fishing can be a traditional trade passed down from generation to generation. Most commercial fishing is based in towns built around the fishing industry; regulation changes can impact an entire town's economy. Cuts in harvest quotas can have adverse effects on the ability of fishermen to compete with the tourism industry.

PERFORMANCE

The biomass of global fish stocks has been allowed to run down. This biomass is now diminished to the point where it is no longer possible to sustainably catch the amount of fish that could be caught. According to a 2008 UN report, titled *The Sunken Billions: The Economic Justification for Fisheries Reform*, the world's fishing fleets incur a "$US 50 billion annual economic loss" through depleted stocks and poor fisheries management. The report, produced jointly by the World Bank and the UN Food and Agriculture Organization (FAO), asserts that half the world's fishing fleet could be scrapped with no change in catch.

By improving governance of marine fisheries, society could capture a substantial part of this $50 billion annual economic loss. Through comprehensive reform, the fisheries sector could become a basis for economic growth and the creation of alternative livelihoods in many countries. At the same time, a nation's natural capital in the form of fish stocks could be greatly increased and the negative impacts of the fisheries on the marine environment reduced.

The most prominent failure of fisheries management in recent times has perhaps been the events that lead to the collapse of the northern cod fisheries. More recently, the International Consortium of Investigative Journalists produced a series of journalistic investigations called *Looting the seas*. These detail investigations into the black market for bluefin tuna, the subsidies propping up the Spanish fishing industry, and the overfishing of the Chilean jack mackerel.

ENDOLITH

An **endolith** is an organism (archaeum, bacterium, fungus, lichen, algae or amoeba) that lives inside rock, coral, animal shells, or in the pores between mineral grains of a rock. Many are extremophiles, living in places previously thought inhospitable to life. They are of particular interest to astrobiologists, who theorize that endolithic environments on Mars and other planets constitute potential refugia for extraterrestrial microbial communities.

Subdefinitions

The term "endolith", which defines an organism that colonizes the interior of any kind of rock, has been further classified into three subclasses:

Chasmoendolith

colonizes fissures and cracks in the rock (*chasm* = cleft)

Cryptoendolith

colonizes structural cavities within porous rocks, including spaces produced and vacated by euendoliths (*crypto* = hidden)

Euendolith

penetrates actively into the interior of rocks forming tunnels that conform with the shape of its body, rock boring organism (*eu* = good, true)

Environment

Endoliths have been found in rock down to a depth of 3 kilometres (1.9 mi), though it is unknown if that is their limit (due to the cost involved in digging so deep). The main threat to their survival seems not to result from the pressure at such depth, but from the increased temperature. Judging from hyperthermophile organisms, the temperature limit is at about 120 °C (Strain 121 can reproduce at 121 °C), which limits the possible depth to 4-4.5 km below the continental crust, and 7 or 7.5 km below the ocean floor. Endolithic organisms have also been found in surface rocks in regions of low humidity (hypolith) and low temperature (psychrophile), including the Dry Valleys and permafrost of Antarctica, the Alps, and the Rocky Mountains.

Survival

Endoliths can survive by feeding on traces of iron, potassium, or sulfur. (See lithotroph.) Whether they metabolize these directly from the surrounding rock, or rather excrete an acid to dissolve them first, remains to be seen. The Ocean Drilling Program found microscopic trails in basalt from the Atlantic, Indian, and Pacific oceans that contain DNA. Photosynthetic endoliths have also been discovered.

As water and nutrients are rather sparse in the environment of the endolith, they have a very slow reproduction cycle. Early data suggests that some only engage in cell division once every hundred years. In August 2013 researchers reported evidence of endoliths in the ocean floor with a generation time of 10,000 years. Most of their energy is spent repairing cell damage caused by cosmic rays or racemization, and very little is available for reproduction or growth. It is thought that they weather long ice ages in this fashion, emerging when the temperature in the area warms.

Slime

As most endoliths are autotrophs, they can generate organic compounds essential for their survival on their own from inorganic matter. Some endoliths have specialized in feeding on their autotroph relatives. The micro-biotope where these different endolithic species live together has been called a *Subsurface Lithotrophic Microbial Ecosystem* (*SLiME*).

Endolithic Fungi and algae in Marine Ecosystems

Only limited research has been done concerning the distribution of marine endolithic fungi and its diversity even though there is a probability that endolithic fungi could perhaps play an important role in the health of coral reefs.

Endolithic fungi have been discovered in shells as early as the year 1889 by Edouard Bornet and Charles Flahault. These two French phycologists specifically provided descriptions for two fungi: *Ostracoblabe implexis* and *Lithopythium gangliiforme*. Discovery of endolithic fungi, such as *Dodgella priscus* and *Conchyliastrum*, has also been made in the beach sand of Australia by George Zembrowski. Findings have also been made in coral reefs and have been found to be, at times, beneficial to their coral hosts.

In the wake of worldwide coral bleaching, studies have suggested that the endolithic algae located in the skeleton of the coral may be aiding the survival of coral species by providing an alternative source of energy. Although the role that endolithic Fungi is important in coral reefs, it is often overlooked because much research is focused on the effects of coral bleaching as well as the relationships between coelenterate and endosymbiotic symbiodinia.

According to a study done by Astrid Gunther endoliths were also found in the island of Cozumel (Mexico). The endoliths found there not only included algae and fungi but also included cyanobacteria, sponges as well as many other microborers.

Endolithic Parasitism

Until the 1990s phototrophic endoliths were thought as somewhat benign but since then evidence has surfaced that phototrophic endoliths (primarily cyanobacteria) have infested 50 to 80% of midshore populations of the mussel species *Perna perna* located in South Africa.The infestation of Phototrophic endoliths resulted in lethal and sub-lethal effects such as the decrease in strength of the mussel shells. Although the rate of thickening of the shells were faster in more infested areas it is not rapid enough to combat the degradation of the mussel shells.

Endolithic fungi and the Mass Extinction of Cretaceous Dinosaurs

Evidence of endolithic fungi were discovered within dinosaur eggshell found in central China. The endolithic fungi that was found within the eggshells were characterized as being "needle-like, ribbon-like, and silk-like."

Fungus is seldom fossilized and even when it is preserved it can be difficult to distinguish endolithic hyphae from endolithic cyanobacteria and algae. Endolithic microbes can, however, be distinguished based on their distribution, ecology, and morphology. According to a 2008 study, the endolithic fungi that formed on the eggshells would have resulted in the abnormal incubation of the eggs and eventually led to the contribution to the mass extinction of these dinosaurs. It may also have led to the preservation of dinosaur eggs, including some that contained embryos.

REFERENCES

1. FAO (1997) Fisheries Management Section 1.2, Technical Guidelines for Responsible Fisheries. FAO, Rome. ISBN 92-5-103962-3
2. Elements of Ecology - Thomas M. Smith, Robert Leo Smith
3. Meredith P (2009) Te hī ika – Māori fishing - Traditional practices Te Ara - the Encyclopedia of New Zealand. Updated 2 March 2009. Retrieved 22 February 2011.
4. Christy FT (1983) Territorial Use Rights in Fisheries: Definitions and Conditions FAO Fisheries, Technical Paper No. 227, Rome. ISBN 978-92-5-101269-7.
5. Hannesson R, Salvanes JG and Squires A (2008) Technological change and the Tragedy of the Commons: The Lofoten Fishery over Hundred and Thirty YearsInstitutt for Samfunnsøkonomi. Discussion paper SAM 5 2008.
6. Beverton & Holt 1957
7. Beverton 1992
8. Prince Charles calls for greater sustainability in fisheries. London Mercury. Retrieved 13 July 2014.
9. FAO, Rome (2009) A Fishery Managers GuidebookEds. Cochrane KL and Garcia S. ISBN 978-1-4051-7085-7
10. Duzgunes, E; Erdogan, N (2008). Fisheries Management in the Black Sea Countries (PDF). Turkish Journal of Fisheries and Aquatic Sciences. 8: 181–192.
11. Code of Conduct for Responsible Fisheries
12. Pitcher, Kalikoski & Pramod 2006
13. Sustainable Fishery system, Charles, A. (ed) (2001). Oxford: Blackwell science. p. 95.
14. Costello, Christopher; Gaines, Steven D.; Lynham, John (2008). Can Catch Shares Prevent Fisheries Collapse?.

15. Debora MacKenzie New Scientist: Guaranteed fish quotas halt commercial free-for-all
16. A Rising Tide: Scientists find proof that privatising fishing stocks can avert a disaster The Economist, 18 Sept, 2008.
17. New study offers solution to global fisheries collapseEureka alert.
18. Soliman, Adam (2014) Using individual transferable quotas (ITQs) to achieve social policy objectives: A proposed intervention, Marine Policy, 45C: 76–81.
19. Soliman, Adam (2014) Individual transferable quotas in world fisheries: Addressing legal and rights-based issuesOcean and Coastal Management Journal, 87: 102–113.
20. FAO (2003) Fisheries management: The ecosystem approach to fisheries Annex 2. Principles of relevance to an ecosystem approach to fisheries (EAF)], Page 85. Rome. ISBN 92-5-104897-5.
21. Dayton PK (1998) "Reversal of the burden of proof in fisheries management" Science, 279(5352): 821– 822.
22. Shertzer KW and Prager MH (2007) Delay in fishery management: diminished yield, longer rebuilding, and increased probability of stock collapse ICES J. Mar. Sci.64: 149–159.
23. Gucinski, Lackey, and Spence (1990)
24. Hlohowskyj, Brody, and Lackey, (1996)
25. Wilderbuera, Thomas K and Zhang, Chang Ik (1999) Evaluation of the population dynamics and yield characteristics of Alaska plaice, Pleuronectes quadrituberculatus, in the eastern Bering Sea. Fisheries Research. Volume 41, Issue 2.
26. Richard W Zabel, Chris J Harvey, Steven L Katz, Thomas P Good, Phillip S Levin (2003) Ecologically Sustainable Yield.American Scientist, March–April.
27. Caswell, H. 2001. Matrix population models: Construction, analysis and interpretation, 2nd Edition. Sinauer Associates, Sunderland, Massachusetts. ISBN 0-87893-096-5.
28. Larkin, PA (1977). An epitaph for the concept of maximum sustained yield (PDF). Transactions of the American Fisheries Society. 106: 1–11. doi:10.1577/1548-8659(1977)106<1:AEFTCO>2.0.CO;2.
29. Walters, C; Maguire, J (1996). Lessons for stock assessment from the northern cod collapse. Reviews in Fish Biology and Fisheries. 6: 125–137. doi:10.1007/bf00182340.
30. Pitcher TJ and Pauly D (1998) Rebuilding ecosystems, not sustainability, as the proper goal of fishery managementPages 311-325 in T Pitcher, D Pauly and P Hart, Reinventing Fisheries Management, Chapman & Hall.
31. University of Liverpool (2006). "Marine Ecologists To Help Rebuild Decreasing Fish Stocks" ScienceDaily.
32. FAO: Fisheries governance: The ecosystem approach to fisheries management

Rome. Updated 27 May 2005. Accessed 27 November 2009.

33. Francis RC, Hixon MA, Clarke ME, Murawski SA, and Ralston S (2007) Ten commandments for ecosystem-based fisheries scientists Proceedings of Coastal Zone 07, Portland, Oregon. Download
34. AAAS (2005) New Science Sheds Light on Rebuilding Fisheries
35. Adkison advocates increased fisheries data gathering. University of Alaska Fairbanks. 2007.
36. Fishery statistics: Reliability and policy implications
37. Mahon & McConney 2004
38. Hilborn, R (2007). Managing fisheries is managing people: what has been learned?. *Fish and Fisheries.* 8 *(4): 285–296. doi:10.1111/j.1467-2979.2007.00263_2.x.*
39. Elder 2006
40. Arnason, Kelleher & Willmann 2008
41. Looting the Seas iWatch News, 17 March 2012.
42. Wierzchos, J.; Camara, B.; De Los Rios, A.; Davila, A. F.; Sanchaz Almazo, M.; Artieda, O.; Wierzchos, K.; Gomez-Silva, B.; McKay, C.; Ascaso, C. (2011). "Microbial colonization of Ca-sulfate crusts in the hyperarid core of the Atacama Desert: Implications for the search for life on Mars". Geobiology. 9 *(1): 44–60. doi:10.1111/j.1472-4669.2010.00254.x.*
43. Chang, Kenneth (12 September 2016). Visions of Life on Mars in Earths Depths. New York Times. Retrieved 12 September 2016.
44. Stjepko Golubic; E.Imre Friedmann; Jürgen Schneider (June 1981). The lithobiotic ecological niche, with special reference to microorganisms. Journal of Sedimentary Research. 51 *(2): 475–478.*
45. Schultz, Steven (13 December 1999). Two miles underground. Princeton Weekly Bulletin. Archived from the original on 13 January 2016. — Gold mines present "ideal environment" for geologists studying subsurface microbes
46. Looking for life in all the wrong places — research on cryptoendoliths Discover, May 1997 by Will Hively
47. Microbial Diversity of Cryptoendolithic Communities from the McMurdo Dry Valleys, Antarctica
48. Horath, Thomas; Bachofen, Reinhard (August 2009). Molecular Characterization of an Endolithic Microbial Community in Dolomite Rock in the Central Alps (Switzerland). Microbial Ecology. 58 *(2): 290–306.* doi:10.1007/s00248-008-9483-7.
49. Walker JJ, Spear JR, Pace NR (2005) Geobiology of a microbial endolithic community in the Yellowstone geothermal environment. Nature 434:1011–1014
50. Walker JJ, Pace NR (2007) Phylogenetic composition of Rocky Mountain endolithic microbial ecosystems. Appl Environ Microbiol 73:3497–3504
51. Mullen, Leslie. Glass Munchers Under the Sea. NASA Astrobiology Institute. Archived from the original on 20 February 2013.

52. Lysnes, Kristine; Torsvik, Terje; Thorseth, Ingunn H.; Pedersen, Rolf B. (2004). Microbial Populations in Ocean Floor Basalt: Results from ODP Leg 187 (PDF). Proc ODP Sci Results. 187: 1–27.
53. Wierzchos, J., Ascaso, C., and McKay., C. P.(2006). Astrobiology, 6(3): 415-422. doi:10.1089/ast.2006.6.415.
54. Yirka, Bob (29 August 2013). Soil beneath ocean found to harbor long lived bacteria, fungi and viruses. Phys.org. Archived from the original on 15 January 2016.
55. Frequently Requested Information about the SLiME Hypothesis. Archived from the original on 30 September 2006.
56. Golubic, Stjepko, Gudrun Radtke, and Therese Le Campion-Alsumard. Endolithic fungi in marine ecosystems. Trends in microbiology 13.5 (2005): 229-235.
57. Fine, M., and Y. Loya. Endolithic Algae: An Alternative Source of Photoassimilates during Coral Bleaching. Proceedings of the Royal Society B: Biological Sciences 269.1497 (2002): 1205-210. Web.
58. Günther, Dipl Biol Astrid. Distribution and bathymetric zonation of shell-boring endoliths in recent reef and shelf environments: Cozumel, Yucatan (Mexico). Facies 22.1 (1990): 233-261.
59. Kaehler, S., and C. D. McQuaid. Lethal and sub-lethal effects of phototrophic endoliths attacking the shell of the intertidal mussel Perna perna. Marine Biology 135.3 (1999): 497-503.
60. Gong, YiMing, Ran Xu, and Bi Hu. Endolithic fungi: A possible killer for the mass extinction of Cretaceous dinosaurs. Science in China Series D: Earth Sciences 51.6 (2008): 801-807.

CHAPTER 9

Industrial Ecology

Chapter Contents

Industrial Ecology is the study of industrial systems aimed at identifying and implementing strategies that reduce their environmental impact. Industries, such as manufacturing and energy plants, extract raw materials and natural resources from the earth and transform them into products and services that meet the demands of the population.

Industrial ecology was developed as a way to better understand the impact industry has on the environment. Therefore, industrial ecologists, or those who study industrial ecology, work to understand the industrial systems that are in place and find ways to use fewer natural resources and find new uses for waste materials or byproducts.

Industrial ecologists approach problems with the mindset that industrial systems can become more efficient and sustainable if the problems are looked at from all directions. Therefore, it is right to think of industrial ecology as a multidisciplinary field that combines aspects of economics, engineering, sociology, technology and environmental science.

INTRODUCTION

Industrial ecology (IE) is the study of material and energy flows through industrial systems. The global industrial economy can be modelled as a network of industrial processes that extract resources from the Earth and transform those resources into commodities which can be bought and sold to meet the needs of humanity. Industrial ecology seeks to quantify the material flows and document the industrial processes that make modern society function. Industrial ecologists are often concerned with the impacts that industrial activities have on the environment, with use of the planet›s supply of natural resources, and with problems of waste disposal. Industrial ecology is a young but growing multidisciplinary field of research which combines aspects of engineering, economics, sociology, toxicology and the natural sciences.

Industrial ecology has been defined as a systems-based, multidisciplinary discourse that seeks to understand emergent behaviour of complex integrated human/natural systems. The field approaches issues of sustainability by examining problems from multiple perspectives, usually involving aspects of sociology, the environment, economy and technology. The name comes from the idea that the analogy of natural systems should be used as an aid in understanding how to design sustainable industrial systems.

OVERVIEW

Example of Industrial Symbiosis. Waste steam from a waste incinerator (right) is piped to an ethanol plant (left) where it is used as in input to their production process.

Industrial ecology is concerned with the shifting of industrial process from linear (open loop) systems, in which resource and capital investments move through the system to become waste, to a closed loop system where wastes can become inputs for new processes.

Much of the research focuses on the following areas:

- material and energy flow studies ("industrial metabolism")
- dematerialization and decarbonization
- technological change and the environment
- life-cycle planning, design and assessment

- design for the environment ("eco-design")
- extended producer responsibility ("product stewardship")
- eco-industrial parks ("industrial symbiosis")
- product-oriented environmental policy
- eco-efficiency

Industrial ecology seeks to understand the way in which industrial systems (for example a factory, an ecoregion, or national or global economy) interact with the biosphere. Natural ecosystems provide a metaphor for understanding how different parts of industrial systems interact with one another, in an "ecosystem" based on resources and infrastructural capital rather than on natural capital. It seeks to exploit the idea that natural systems do not have waste in them to inspire sustainable design.

Along with more general energy conservation and material conservation goals, and redefining commodity markets and product stewardship relations strictly as a service economy, industrial ecology is one of the four objectives of Natural Capitalism. This strategy discourages forms of amoral purchasing arising from ignorance of what goes on at a distance and implies a political economy that values natural capital highly and relies on more instructional capital to design and maintain each unique industrial ecology.

HISTORY

View of Kalundborg Eco-industrial Park

Industrial ecology was popularized in 1989 in a *Scientific American* article by Robert Frosch and Nicholas E. Gallopoulos. Frosch and Gallopoulos› vision was why would not our industrial system behave like an ecosystem, where the wastes of a species may be resource to another species? Why would not the outputs of an industry be the inputs of another, thus reducing use of raw materials, pollution, and saving on waste treatment? A notable example resides in a Danish industrial park in the city of Kalundborg. Here several linkages of byproducts and waste heat can be found between numerous entities such as a large power plant, an oil refinery, a pharmaceutical plant, a plasterboard factory, an enzyme manufacturer, a waste company and the city itself.

The scientific field Industrial Ecology has grown quickly in recent years. The Journal of Industrial Ecology (since 1997), the International Society for Industrial Ecology (since 2001), and the journal Progress in Industrial Ecology (since 2004) give Industrial Ecology a strong and dynamic position in the international scientific community. Industrial Ecology principles are also emerging in various policy realms such as the concept of the Circular Economy that is being promoted in China. Although the definition of the Circular Economy has yet to be formalized, generally the focus is on strategies such as creating a circular flow of materials, and cascading energy flows. An example of this would be using waste heat from one process to run another process that requires a lower temperature. The hope is that strategy such as this will create a more efficient economy with fewer pollutants and other unwanted by-products.

PRINCIPLES

One of the central principles of Industrial Ecology is the view that societal and technological systems are bounded within the biosphere, and do not exist outside of it. Ecology is used as a *metaphor* due to the observation that natural systems reuse materials and have a largely closed loop cycling of nutrients. Industrial Ecology approaches problems with the hypothesis that by using similar principles as *natural systems, industrial systems* can be improved to reduce their impact on the natural environment as well. The table shows the general metaphor.

Biosphere	**Technosphere**
• Environment	• Market
• Organism	• Company
• Natural Product	• Industrial Product
• Natural Selection	• Competition
• Ecosystem	• Eco-Industrial Park
• Ecological Niche	• Market Niche
• Anabolism / Catabolism	• Manufacturing / Waste Management
• Mutation and Selection	• Design for Environment
• Succession	• Economic Growth
• Adaptation	• Innovation
• Food Web	• Product Life Cycle

The Kalundborg industrial park is located in Denmark. This industrial park is special because companies reuse each other's waste (which then becomes by-products). For example, the Energy E2 Asnæs Power Station produces gypsum as a by-product of the electricity generation process; this gypsum becomes a resource for the BPB Gyproc A/S which produces plasterboards. This is one example of a system inspired by the biosphere-technosphere metaphor: in ecosystems, the waste from one organism is used as inputs to other organisms; in industrial systems, waste from a company is used as a resource by others.

Apart from the direct benefit of incorporating waste into the loop, the use of an eco-industrial park can be a means of making renewable energy generating plants, like Solar PV, more economical and environmentally friendly. In essence, this assists the growth of the renewable energy industry and the environmental benefits that come with replacing fossil-fuels.

IE examines societal issues and their relationship with both technical systems and the environment. Through this *holistic view* , IE recognizes that solving problems must involve understanding the connections that exist between these systems, various aspects cannot be viewed in isolation. Often changes in one part of the overall system can propagate and cause changes in another part. Thus, you can only understand a problem if you look at its parts in relation to the whole. Based on this framework, IE looks at environmental issues with a *systems thinking* approach.

Take a city for instance. A city can be divided into commercial areas, residential areas, offices, services, infrastructures, and so forth. These are all sub-systems of the big city' system. Problems can emerge in one sub-system, but the solution has to be global. Let's say the price of housing is rising dramatically because there is too high a demand for housing. One solution would be to build new houses, but this will lead to more people living in the city, leading to the need for more infrastructure like roads, schools, more supermarkets, etc. This system is a simplified interpretation of reality whose behaviors can be 'predicted'.

In many cases, the systems IE deals with are complex systems. Complexity makes it difficult to understand the behavior of the system and may lead to rebound effects. Due to unforeseen behavioral change of users or consumers, a measure taken to improve environmental performance does not lead to any improvement or may even worsen the situation.

Moreover, *life cycle thinking* is also a very important principle in industrial ecology. It implies that all environmental impacts caused by a product, system, or project during its life cycle are taken into account. In this context life cycle includes

- Raw material extraction
- Material processing
- Manufacture
- Use
- Maintenance
- Disposal

The transport necessary between these stages is also taken into account as well as, if relevant, extra stages such as reuse, remanufacture, and recycle. Adopting a life cycle approach is essential to avoid shifting environmental impacts from one life cycle stage to another. This is commonly referred to as problem shifting. For instance, during the re-design of a product, one can choose to reduce its weight, thereby decreasing use of resources. However, it is possible that the lighter materials used in the new product will be more difficult to dispose of. The environmental impacts of the product gained during the extraction phase are shifted to the disposal phase. Overall environmental improvements are thus null.

A final and important principle of IE is its *integrated approach* or *multidisciplinarity*. IE takes into account three different disciplines: social sciences (including economics), technical sciences and environmental sciences. The challenge is to merge them into a single approach.

TOOLS

People	Planet	Profit	Modeling
• Stakeholder analysis • Strength Weakness Opportunities Threats Analysis (SWOT Analysis) • Ecolabelling • ISO 14000 • Environmental management system (EMS) • Integrated chain management (ICM) • Technology assessment	• Environmental impact assessment (EIA) • Input-output analysis (IOA) • Life-cycle assessment (LCA) • Material flow analysis (MFA) • Substance flow analysis (SFA) • MET Matrix	• Cost benefit analysis (CBA) • Full cost accounting (FCA) • Life cycle costing (LCC)	• Stock and flow analysis • Agent based modeling

FUTURE DIRECTIONS

The ecosystem metaphor popularized by Frosch and Gallopoulos has been a valuable creative tool for helping researchers look for novel solutions to difficult problems. Recently, it has been pointed out that this metaphor is based largely on a model of classical ecology, and that advancements in understanding ecology based on complexity science have been made by researchers such as C. S. Holling, James J. Kay, and further advanced in terms of contemporary ecology by others. For industrial ecology, this may mean a shift from a more mechanistic view of systems, to one where sustainability is viewed as an emergent property of a complex system. To explore this further, several researchers are working with agent based modeling techniques .

Exergy analysis is performed in the field of industrial ecology to use energy more efficiently. The term *exergy* was coined by Zoran Rant in 1956, but the concept was developed by J. Willard Gibbs. In recent decades, utilization of exergy has spread outside of physics and engineering to the fields of industrial ecology, ecological economics, systems ecology, and energetics.

MACROECOLOGY

Macroecology is the subfield of ecology that deals with the study of relationships between organisms and their environment at large spatial scales to characterise and explain statistical patterns of abundance, distribution and diversity. The term was coined by James Brown of the University of New Mexico and Brian Maurer of Michigan State University in a 1989 paper in *Science*.

Macroecology approaches the idea of studying ecosystems using a "top down" approach. It seeks understanding through the study of the properties of the system as a whole; Kevin Gaston and Tim Blackburn make the analogy to seeing the forest for the trees.

Macroecology examines how global development in climate change affect wildlife populations. Classic ecological questions amenable to study through the techniques of macroecology include questions of species richness, latitudinal gradients in species diversity, the species-area curve, range size, body size, and species abundance. For example, the relationship between abundance and range size (why species that maintain large local population sizes tend to be widely distributed, while species that are less abundant tend to have restricted ranges) has received much attention.

REFERENCES

1. Allenby, Brad (2006). The ontologies of industrial ecology (PDF). Progress in Industrial Ecology. Inderscience Enterprises Ltd. **3** *(1/2): 28–40. doi:10.1504/PIE.2006.010039.*
2. Frosch, R.A.; Gallopoulos, N.E. (1989). Strategies for Manufacturing. Scientific American. **261** *(3): 144–152.* doi:10.1038/scientificamerican0989-144.
3. International Society for Industrial Ecology | History. Retrieved January 8, 2009.
4. The Kalundborg Centre for Industrial Symbiosis. Retrieved 2007. Check date values in: |access-date= (help)
5. Yuan, Z.; Bi, J.; Moriguichi, Y. The Circular Economy: A New Development Strategy in China. Journal of Industrial Ecology. **10** *(1–2): 4–8.* doi:10.1162/108819806775545321.
6. Pearce, J. M. (2008). Industrial Symbiosis for Very Large Scale Photovoltaic Manufacturing. Renewable Energy. **33** *(5): 1101–1108. doi:10.1016/j.renene.2007.07.002.*
7. Kay, J.J. (2002). Kibert, C.; Sendzimir, J.; Guy, B., eds. On Complexity Theory,

Exergy and Industrial Ecology: Some Implications for Construction Ecology (PDF). Construction Ecology: Nature as the Basis for Green Buildings. Spon Press: 72–107

8. Levine, S. H. (2003), Comparing Products and Production in Ecological and Industrial Systems. Journal of Industrial Ecology 7(2): 33-42.

9. Nielsen, Søren Nors (2007). What has modern ecosystem theory to offer to cleaner production, industrial ecology and society? The views of an ecologist. Journal of Cleaner Production. **15** *(17): 1639–1653. doi:10.1016/j.jclepro.2006.08.008.*

10. Ashton, W. S. (2009), The Structure, Function, and Evolution of a Regional Industrial Ecosystem. Journal of Industrial Ecology 13(2): 228.

11. Jensen, P. D. (2011), Reinterpreting Industrial Ecology. Journal of Industrial Ecology 15(5):680-692.

12. Ehrenfeld, John (2004). Can Industrial Ecology be the Science of Sustainability?. *Journal of Industrial Ecology.* **8** *(1–2): 1–3.* doi:10.1162/1088198041269364.

13. Ehrenfeld, John (2007). Would Industrial Ecology Exist without Sustainability in the Background?. *Journal of Industrial Ecology.* **11** *(1).*

14. Axtell, R.L.; Andrews, C.J.; Small, M.J. (2002). Agent-Based Modeling and Industrial Ecology. Journal of Industrial Ecology. **5** *(4): 10–13.* doi:10.1162/10881980160084006.

15. Kraines, S.; Wallace, D. (2006). Applying Agent-based Simulation in Industrial Ecology. Journal of Industrial Ecology. **10** *(1–2): 15–18.* doi:10.1162/108819806775545376.

16. Wall, Göran. Exergy - a useful concept

CHAPTER 10

Natural Environment

Chapter Contents

The natural environment is God's gift and provision for the care of his people. It provides healthy food, shelter, clothing, medicines, beauty, clean water, clean air and many other benefits for people to use and enjoy. The natural environment is also the world's largest employer. More people all over the world make their living from environmental resources than by any other means.

World Vision works with communities to restore environments that have been harmed, protect healthy environments, and systematically build good environments to make them progressively stronger and healthier.

INTRODUCTION

The natural environment encompasses all living and non-living things occurring naturally. The term is most often applied to the Earth or some part of Earth. This environment encompasses the interaction of all living species, climate, weather, and natural resources that affect human survival and economic activity. The concept of the *natural environment* can be distinguished by components:

Land management has preserved the natural characteristics of Hopetoun Falls, Australia while allowing ample access for visitors.

Satellite image of the Sahara desert; the world's largest hot desert and third-largest desert after the polar deserts.

- Complete ecological units that function as natural systems without massive civilized human intervention, including all vegetation, microorganisms, soil, rocks, atmosphere, and natural phenomena that occur within their boundaries and their nature
- Universal natural resources and physical phenomena that lack clear-cut boundaries, such as air, water, and climate, as well as energy, radiation, electric charge, and magnetism, not originating from civilized human activity

In contrast to the natural environment is the built environment. In such areas where man has fundamentally transformed landscapes such as urban settings and agricultural land conversion, the natural environment is greatly modified into a simplified human environment. Even acts which seem less extreme, such as building a mud hut or a photovoltaic system in the desert, modify the natural environment into an artificial one. Though many animals build things to provide a better environment for themselves, they are not human, hence beaver dams and the works of Mound-building termites are thought of as natural.

It is difficult to find *absolutely natural* environments on Earth, and naturalness usually varies in a continuum, from 100% natural in one extreme to 0% natural in the other. More precisely, we can consider the different aspects or components of an environment, and see that their degree of naturalness is not uniform. If, for instance, in an agricultural field, the mineralogic composition and the structure of its soil are similar to those of an undisturbed forest soil, but the structure is quite different.

Natural environment is often used as a synonym for habitat. For instance, when we say that the natural environment of giraffes is the savanna.

COMPOSITION

A volcanic fissure and lava channel.

Earth science

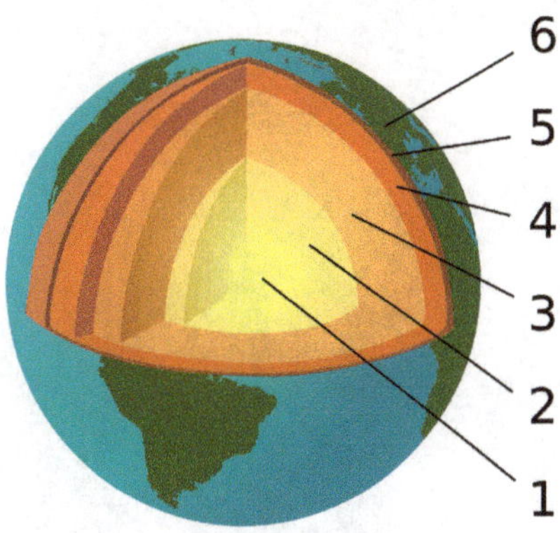

Earth's layered structure. (1) inner core; (2) outer core; (3) lower mantle; (4) upper mantle; (5) lithosphere; (6) crust

Earth science generally recognizes 4 spheres, the lithosphere, the hydrosphere, the atmosphere, and the biosphere as correspondent to rocks, water, air, and life respectively.

Some scientists include, as part of the spheres of the Earth, the cryosphere (corresponding to ice) as a distinct portion of the hydrosphere, as well as the pedosphere (corresponding to soil) as an active and intermixed sphere. Earth science (also known as geoscience, the geosciences or the Earth Sciences), is an all-embracing term for the sciences related to the planet Earth. There are four major disciplines in earth sciences, namely geography, geology, geophysics and geodesy. These major disciplines use physics, chemistry, biology, chronology and mathematics to build a qualitative and quantitative understanding of the principal areas or *spheres* of Earth.

GEOLOGICAL ACTIVITY

The Earth's crust, or lithosphere, is the outermost solid surface of the planet and is chemically and mechanically different from underlying mantle. It has been generated greatly by igneous processes in which magma cools and solidifies to form solid rock. Beneath the lithosphere lies the mantle which is heated by the decay of radioactive elements. The mantle though solid is in a state of rheic convection. This convection process causes the lithospheric plates to move, albeit slowly. The resulting process is known as plate tectonics. Volcanoes result primarily from the melting of subducted crust material or of rising mantle at mid-ocean ridges and mantle plumes.

Water on Earth

Coral reefs have significant marine biodiversity.

Oceans

An ocean is a major body of saline water, and a component of the hydrosphere. Approximately 71% of the Earth›s surface (an area of some 362 million square kilometers) is covered by ocean, a continuous body of water that is customarily divided into several principal oceans and smaller seas. More than half of this area is over 3,000 meters (9,800 ft) deep. Average oceanic salinity is around 35 parts per thousand (ppt) (3.5%), and nearly all seawater has a salinity in the range of 30 to 38 ppt. Though generally recognized as several ‹separate› oceans, these waters comprise one global, interconnected body of salt water often referred to as the World Ocean or global ocean. The deep seabeds are more than half the Earth's surface, and are among the least-modified natural environments. The major oceanic divisions are defined in part by the continents, various archipelagos, and other criteria: these divisions are (in descending order of size) the Pacific Ocean, the Atlantic Ocean, the Indian Ocean, the Southern Ocean and the Arctic Ocean.

Rivers

The Columbia River, along the border of the U.S. states of Oregon and Washington.

A rocky stream in the U.S. state of Hawaii

A river is a natural watercourse, usually freshwater, flowing toward an ocean, a lake, a sea or another river. In a few cases, a river simply flows into the ground and dries up completely before reaching another body of water. Small rivers may also be termed by several other names, including stream, creek and brook. In the United States a river is generally classified as a watercourse more than 60 feet (18 metres) wide. The water in a river is usually in a channel, made up of a stream bed between banks. In larger rivers there is also a wider floodplain shaped by waters over-topping the channel. Flood plains may be very wide in relation to the size of the river channel. Rivers are a part of the hydrological cycle. Water within a river is generally collected from precipitation through surface runoff, groundwater recharge, springs, and the release of water stored in glaciers and snowpacks.

Streams

A stream is a flowing body of water with a current, confined within a bed and stream banks. Streams play an important corridor role in connecting fragmented habitats and thus in conserving biodiversity. The study of streams and waterways in general is known as *surface hydrology.* Types of streams include creeks, tributaries, which do not reach an ocean and connect with another stream or river, brooks, which are typically small streams and sometimes sourced from a spring or seep and tidal inlets.

Lakes

Lácar Lake, of glacial origin, in the province of Neuquén, Argentina.

A lake (from Latin *lacus*) is a terrain feature, a body of water that is localized to the bottom of basin. A body of water is considered a lake when it is inland, is not part of an ocean, and is larger and deeper than a pond.

A swamp area in Everglades National Park, Florida, US.

Natural lakes on Earth are generally found in mountainous areas, rift zones, and areas with ongoing or recent glaciation. Other lakes are found in endorheic basins or along the courses of mature rivers. In some parts of the world, there are many lakes because of chaotic drainage patterns left over from the last Ice Age. All lakes are temporary over geologic time scales, as they will slowly fill in with sediments or spill out of the basin containing them.

Ponds

A pond is a body of standing water, either natural or man-made, that is usually smaller than a lake. A wide variety of man-made bodies of water are classified as ponds, including water gardens designed for aesthetic ornamentation, fish ponds designed for commercial fish breeding, and solar ponds designed to store thermal energy. Ponds and lakes are distinguished from streams via current speed. While currents in streams are easily observed, ponds and lakes possess thermally driven micro-currents and moderate wind driven currents. These features distinguish a pond from many other aquatic terrain features, such as stream pools and tide pools.

ATMOSPHERE, CLIMATE AND WEATHER

Atmospheric gases scatter blue light more than other wavelengths, creating a blue halo when seen from space.

A view of Earth's troposphere from an airplane.

Lightning is an atmospheric discharge of electricity accompanied by thunder, which occurs during thunderstorms and certain other natural conditions.

The atmosphere of the Earth serves as a key factor in sustaining the planetary ecosystem. The thin layer of gases that envelops the Earth is held in place by the planet's gravity. Dry air consists of 78% nitrogen, 21% oxygen, 1% argon and other inert gases, such as carbon dioxide. The remaining gases are often referred to as trace gases, among which are the greenhouse gases such as water vapor, carbon dioxide, methane, nitrous oxide, and ozone. Filtered air includes trace amounts of many other chemical compounds. Air also contains a variable amount of water vapor and suspensions of water droplets and ice crystals seen as clouds. Many natural substances may be present in tiny amounts in an unfiltered air sample, including dust, pollen and spores, sea spray, volcanic ash, and meteoroids. Various industrial pollutants also may be present, such as chlorine (elementary or in compounds), fluorine compounds, elemental mercury, and sulphur compounds such as sulphur dioxide [SO_2].

The ozone layer of the Earth's atmosphere plays an important role in depleting the amount of ultraviolet (UV) radiation that reaches the surface. As DNA is readily damaged by UV light, this serves to protect life at the surface. The atmosphere also retains heat during the night, thereby reducing the daily temperature extremes.

Atmospheric Layers

Principal Layers

Earth's atmosphere can be divided into five main layers. These layers are mainly determined by whether temperature increases or decreases with altitude. From highest to lowest, these layers are:

- Exosphere: The outermost layer of Earth's atmosphere extends from the exobase upward, mainly composed of hydrogen and helium.
- Thermosphere: The top of the thermosphere is the bottom of the exosphere, called the exobase. Its height varies with solar activity and ranges from about 350–800 km (220–500 mi; 1,150,000–2,620,000 ft). The International Space Station orbits in this layer, between 320 and 380 km (200 and 240 mi).
- Mesosphere: The mesosphere extends from the stratopause to 80–85 km (50–53 mi; 262,000–279,000 ft). It is the layer where most meteors burn up upon entering the atmosphere.
- Stratosphere: The stratosphere extends from the tropopause to about 51 km (32 mi; 167,000 ft). The stratopause, which is the boundary between the stratosphere and mesosphere, typically is at 50 to 55 km (31 to 34 mi; 164,000 to 180,000 ft).
- Troposphere: The troposphere begins at the surface and extends to between 7 km (23,000 ft) at the poles and 17 km (56,000 ft) at the equator, with some variation due to weather. The troposphere is mostly heated by transfer of energy from the surface, so on average the lowest part of the troposphere is warmest and temperature decreases with altitude. The tropopause is the boundary between the troposphere and stratosphere.

Other Layers

Within the five principal layers determined by temperature are several layers determined by other properties.

- The ozone layer is contained within the stratosphere. It is mainly located in the lower portion of the stratosphere from about 15–35 km (9.3–21.7 mi; 49,000–115,000 ft), though the thickness varies seasonally and geographically. About 90% of the ozone in our atmosphere is contained in the stratosphere.
- The ionosphere, the part of the atmosphere that is ionized by solar radiation, stretches from 50 to 1,000 km (31 to 621 mi; 160,000 to 3,280,000 ft) and typically overlaps both the exosphere and the thermosphere. It forms the inner edge of the magnetosphere.

- The homosphere and heterosphere: The homosphere includes the troposphere, stratosphere, and mesosphere. The upper part of the heterosphere is composed almost completely of hydrogen, the lightest element.
- The planetary boundary layer is the part of the troposphere that is nearest the Earth's surface and is directly affected by it, mainly through turbulent diffusion.

Effects of Global Warming

The Retreat of glaciers since 1850 of Aletsch Glacier in the Swiss Alps (situation in 1979, 1991 and 2002), due to global warming.

The potential dangers of global warming are being increasingly studied by a wide global consortium of scientists. These scientists are increasingly concerned about the potential long-term effects of global warming on our natural environment and on the planet. Of particular concern is how climate change and global warming caused by anthropogenic, or human-made releases of greenhouse gases, most notably carbon dioxide, can act interactively, and have adverse effects upon the planet, its natural environment and humans› existence. It is clear the planet is warming, and warming rapidly.–This warming is also responsible for the extinction of natural habitats,which in turn leads to a reduction in wildlife population.The most recent report from the Intergovernmental Panel on Climate Change (the group of the leading climate scientists in the world) concluded that the earth will warm anywhere from 2.7 to almost 11 degrees Fahrenheit (1.5 to 6 degrees Celsius) between 1990 and 2100. Efforts have been increasingly focused on the mitigation of greenhouse gases that are causing climatic changes, on developing adaptative strategies to global warming, to assist humans, other animal, and plant species, ecosystems, regions and nations in adjusting to the effects of global warming. Some examples of recent collaboration to address climate change and global warming include:

Another view of the Aletsch Glacier in the Swiss Alps and because of global warming it has been decreasing

- The United Nations Framework Convention Treaty and convention on Climate Change, to stabilize greenhouse gas concentrations in the atmosphere at a level that would prevent dangerous anthropogenic interference with the climate system.
- The Kyoto Protocol, which is the protocol to the international Framework Convention on Climate Change treaty, again with the objective of reducing greenhouse gases in an effort to prevent anthropogenic climate change.
- The Western Climate Initiative, to identify, evaluate, and implement collective and cooperative ways to reduce greenhouse gases in the region, focusing on a market-based cap-and-trade system.

A significantly profound challenge is to identify the natural environmental dynamics in contrast to environmental changes not within natural variances. A common solution is to adapt a static view neglecting natural variances to exist. Methodologically, this view could be defended when looking at processes which change slowly and short time series, while the problem arrives when fast processes turns essential in the object of the study.

Climate

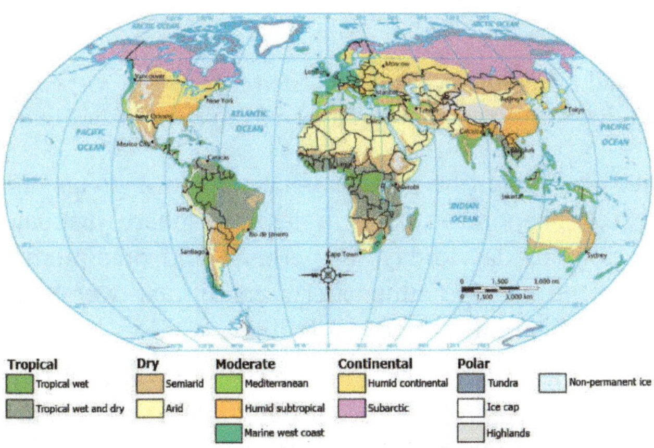

Worldwide climate classifications map

Climate encompasses the statistics of temperature, humidity, atmospheric pressure, wind, rainfall, atmospheric particle count and numerous other meteorological elements in a given region over long periods of time. Climate can be contrasted to weather, which is the present condition of these same elements over periods up to two weeks.

Climates can be classified according to the average and typical ranges of different variables, most commonly temperature and precipitation. The most commonly used classification scheme is the one originally developed by Wladimir Köppen. The Thornthwaite system, in use since 1948, incorporates evapotranspiration in addition to temperature and precipitation information and is used in studying animal species diversity and potential impacts of climate changes.

Weather

Rainbows are optical and meteorological phenomenon that causes a spectrum of light to appear in the sky when the Sun shines onto droplets of moisture in the Earth›s atmosphere.

Weather is a set of all the phenomena occurring in a given atmospheric area at a given time. Most weather phenomena occur in the troposphere, just below the stratosphere. Weather refers, generally, to day-to-day temperature and precipitation activity, whereas climate is the term for the average atmospheric conditions over longer periods of time. When used without qualification, "weather" is understood to be the weather of Earth.

Weather occurs due to density (temperature and moisture) differences between one place and another. These differences can occur due to the sun angle at any particular spot, which varies by latitude from the tropics. The strong temperature contrast between polar and tropical air gives rise to the jet stream. Weather systems in the mid-latitudes, such as extratropical cyclones, are caused by instabilities of the jet stream flow. Because the

Earth›s axis is tilted relative to its orbital plane, sunlight is incident at different angles at different times of the year. On the Earth's surface, temperatures usually range ±40 °C (100 °F to −40 °F) annually. Over thousands of years, changes in the Earth's orbit have affected the amount and distribution of solar energy received by the Earth and influence long-term climate

Surface temperature differences in turn cause pressure differences. Higher altitudes are cooler than lower altitudes due to differences in compressional heating. Weather forecasting is the application of science and technology to predict the state of the atmosphere for a future time and a given location. The atmosphere is a chaotic system, and small changes to one part of the system can grow to have large effects on the system as a whole. Human attempts to control the weather have occurred throughout human history, and there is evidence that civilized human activity such as agriculture and industry has inadvertently modified weather patterns.

LIFE

There are many plant species on the planet.

An example of the many animal species on the Earth.

Evidence suggests that life on Earth has existed for about 3.7 billion years. All known life forms share fundamental molecular mechanisms, and based on these observations, theories on the origin of life attempt to find a mechanism explaining the formation of a primordial single cell organism from which all life originates. There are many different hypotheses regarding the path that might have been taken from simple organic molecules via pre-cellular life to protocells and metabolism.

Although there is no universal agreement on the definition of life, scientists generally accept that the biological manifestation of life is characterized by organization, metabolism, growth, adaptation, response to stimuli and reproduction. Life may also be said to be simply the characteristic state of organisms. In biology, the science of living organisms, life is the condition which distinguishes active organisms from inorganic matter, including the capacity for growth, functional activity and the continual change preceding death.

A diverse variety of living organisms (life forms) can be found in the biosphere on Earth, and properties common to these organisms—plants, animals, fungi, protists, archaea, and bacteria—are a carbon- and water-based cellular form with complex organization and heritable genetic information. Living organisms undergo metabolism, maintain homeostasis, possess a capacity to grow, respond to stimuli, reproduce and, through natural selection, adapt to their environment in successive generations. More complex living organisms can communicate through various means.

ECOSYSTEMS

Rainforests often have a great deal of biodiversity with many plant and animal species. This is the Gambia River in Senegal›s Niokolo-Koba National Park.

An ecosystem (also called as environment) is a natural unit consisting of all plants, animals and micro-organisms (biotic factors) in an area functioning together with all of the non-living physical (abiotic) factors of the environment.

Central to the ecosystem concept is the idea that living organisms are continually engaged in a highly interrelated set of relationships with every other element constituting the environment in which they exist. Eugene Odum, one of the founders of the science of ecology, stated: Any unit that includes all of the organisms (ie: the community) in a given area interacting with the physical environment so that a flow of energy leads to clearly defined trophic structure, biotic diversity, and material cycles (i.e.: exchange of materials between living and nonliving parts) within the system is an ecosystem.

Old-growth forest and a creek on Larch Mountain, in the U.S. state of Oregon.

The human ecosystem concept is then grounded in the deconstruction of the human/nature dichotomy, and the emergent premise that all species are ecologically integrated with each other, as well as with the abiotic constituents of their biotope.

A greater number or variety of species or biological diversity of an ecosystem may contribute to greater resilience of an ecosystem, because there are more species present at a location to respond to change and thus "absorb" or reduce its effects. This reduces the effect before the ecosystem's structure is fundamentally changed to a different state. This is not universally the case and there is no proven relationship between the species diversity of an ecosystem and its ability to provide goods and services on a sustainable level.

The term ecosystem can also pertain to human-made environments, such as human ecosystems and human-influenced ecosystems, and can describe any situation where there is relationship between living organisms and their environment. Fewer areas on the surface of the earth today exist free from human contact, although some genuine wilderness areas continue to exist without any forms of human intervention.

Biomes

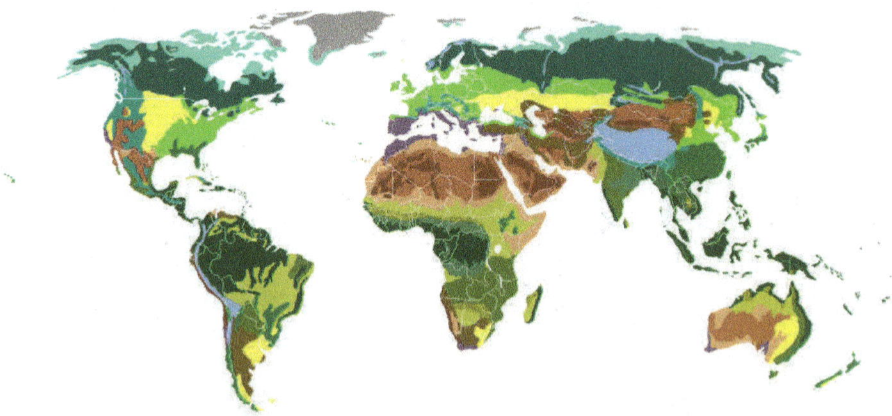

Map of Terrestrial biomes classified by vegetation.

Biomes are terminologically similar to the concept of ecosystems, and are climatically and geographically defined areas of ecologically similar climatic conditions on the Earth, such as communities of plants, animals, and soil organisms, often referred to *as* ecosystems. Biomes are defined on the basis of factors such as plant structures (such as trees, shrubs, and grasses), leaf types (such as broadleaf and needleleaf), plant spacing (forest, woodland, savanna), and climate. Unlike ecozones, biomes are not defined by genetic, taxonomic, or historical similarities. Biomes are often identified with particular patterns of ecological succession and climax vegetation.

BIOGEOCHEMICAL CYCLES

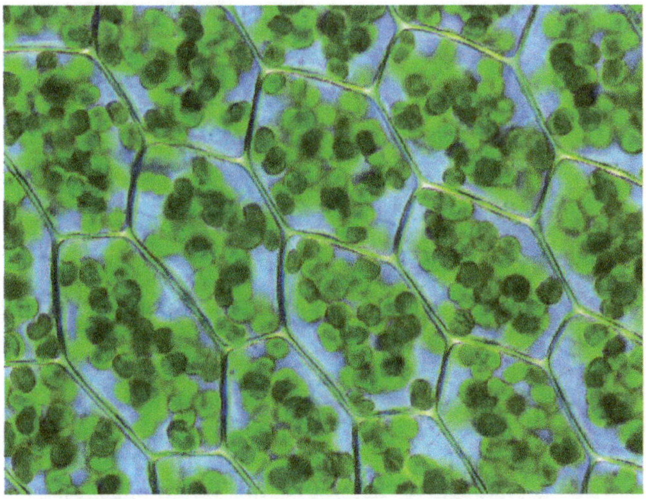

Chloroplasts conduct photosynthesis and are found in plant cells and other eukaryotic organisms. These are Chloroplasts visible in the cells of *Plagiomnium affine* — Many-fruited Thyme-moss.

Global biogeochemical cycles are critical to life, most notably those of water, oxygen, carbon, nitrogen and phosphorus.

1. The nitrogen cycle is the transformation of nitrogen and nitrogen-containing compounds in nature. It is a cycle which includes gaseous components.
2. The water cycle, is the continuous movement of water on, above, and below the surface of the Earth. Water can change states among liquid, vapour, and ice at various places in the water cycle. Although the balance of water on Earth remains fairly constant over time, individual water molecules can come and go.
3. The carbon cycle is the biogeochemical cycle by which carbon is exchanged among the biosphere, pedosphere, geosphere, hydrosphere, and atmosphere of the Earth.
4. The oxygen cycle is the movement of oxygen within and between its three main reservoirs: the atmosphere, the biosphere, and the lithosphere. The main driving factor of the oxygen cycle is photosynthesis, which is responsible for the modern Earth›s atmospheric composition and life.
5. The phosphorus cycle is the movement of phosphorus through the lithosphere, hydrosphere, and biosphere. The atmosphere does not play a significant role in the movements of phosphorus, because phosphorus and phosphorus compounds are usually solids at the typical ranges of temperature and pressure found on Earth.

WILDERNESS

A conifer forest in the Swiss Alps (National Park).

The Ahklun Mountains and the Togiak Wilderness within the Togiak National Wildlife Refuge in the U.S. state of Alaska.

Wilderness is generally defined as a natural environment on Earth that has not been significantly modified by human activity. The WILD Foundation goes into more detail, defining wilderness as: "The most intact, undisturbed wild natural areas left on our planet - those last truly wild places that humans do not control and have not developed with roads, pipelines or other industrial infrastructure." Wilderness areas and protected parks are considered important for the survival of certain species, ecological studies, conservation, solitude, and recreation. Wilderness is deeply valued for cultural, spiritual, moral, and aesthetic reasons. Some nature writers believe wilderness areas are vital for the human spirit and creativity.

The word, wilderness, derives from the notion of wildness; in other words that which is not controllable by humans. The word›s etymology is from the Old English *wildeornes*, which in turn derives from *wildeor* meaning *wild beast* (wild + deor = beast, deer). From this point of view, it is the wildness of a place that makes it a wilderness. The mere presence or activity of people does not disqualify an area from being wilderness. Many ecosystems that are, or have been, inhabited or influenced by activities of people may still be considered wild. This way of looking at wilderness includes areas within which natural processes operate without very noticeable human interference.

Wildlife includes all non-domesticated plants, animals and other organisms. Domesticating wild plant and animal species for human benefit has occurred many times all over the planet, and has a major impact on the environment, both positive and negative. Wildlife can be found in all ecosystems. Deserts, rain forests, plains, and other areas—including the most developed urban sites—all have distinct forms of wildlife. While the term in popular culture usually refers to animals that are untouched by civilized human factors, most scientists agree that wildlife around the world is (now) impacted by human activities.

A view of wilderness in Estonia

CHALLENGES

Before flue-gas desulfurization was installed, the air-polluting emissions from this power plant in New Mexico contained excessive amounts of sulfur dioxide

Amazon Rainforest in Brazil. The tropical rainforests of South America contain the largest diversity of species on Earth, including some that have evolved within the past few hundred thousand years.

It is the common understanding of *natural environment* that underlies environmentalism — a broad political, social, and philosophical movement that advocates various actions and policies in the interest of protecting what nature remains in the natural environment, or restoring or expanding the role of nature in this environment. While true wilderness is increasingly rare, *wild* nature (e.g., unmanaged forests, uncultivated grasslands, wildlife, wildflowers) can be found in many locations previously inhabited by humans.

Goals for the benefit of people and natural systems, commonly expressed by environmental scientists and environmentalists include:

- Elimination of pollution and toxicants in air, water, soil, buildings, manufactured goods, and food.
- Preservation of biodiversity and protection of endangered species.
- Conservation and sustainable use of resources such as water, land, air, energy, raw materials, and natural resources.
- Halting human-induced global warming, which represents pollution, a threat to biodiversity, and a threat to human populations.
- Shifting from fossil fuels to renewable energy in electricity, heating and cooling, and transportation, which addresses pollution, global warming, and sustainability. This may include public transportation and distributed generation, which have benefits for traffic congestion and electric reliability.
- Establishment of nature reserves for recreational purposes and ecosystem preservation.

- Sustainable and less polluting waste management including waste reduction (or even zero waste), reuse, recycling, composting, waste-to-energy, and anaerobic digestion of sewage sludge.

CRITICISM

In some cultures the term environment is meaningless because there is no separation between people and what they view as the natural world, or their surroundings. Specifically in the United States, many native cultures do not recognize the "environment", or see themselves as environmentalists.

REFERENCES

6. Johnson, D. L.; Ambrose, S. H.; Bassett, T. J.; Bowen, M. L.; Crummey, D. E.; Isaacson, J. S.; Johnson, D. N.; Lamb, P.; Saul, M.; Winter-Nelson, A. E. (1997). "Meanings of Environmental Terms". Journal of Environmental Quality. **26** *(3): 581–589. doi:10.2134/jeq1997.00472425002600030002x.*
7. Symons, Donald (1979). The Evolution of Human Sexuality. New York: Oxford University Press. p. 31. ISBN 0-19-502535-0.
8. Earth's Spheres. ©1997-2000. Wheeling Jesuit University/NASA Classroom of the Future. Retrieved November 11, 2007.
9. Wordnet Search: Earth science
10. "[1]". The Columbia Encyclopedia. 2002. New York: Columbia University Press
11. "Distribution of land and water on the planet Archived May 31, 2008, at the Wayback Machine.. UN Atlas of the Oceans ArchivedSeptember 15, 2008, at the Wayback Machine.
12. River {definition} from Merriam-Webster. Accessed February 2010.
13. http://ga.water.usgs.gov/edu/hydrology.html/ Archived August 13, 2009, at the Wayback Machine.
14. Brittanica online. Lake (physical feature). Retrieved 2008-06-25 Thea, Cambodia.. [a Lake is] any relatively large body of slowly moving or standing water that occupies an inland basin of appreciable size. Definitions that precisely distinguish lakes, ponds, swamps, and even rivers and other bodies of nonoceanic water are not established. It may be said, however, that rivers and streams are relatively fast moving; marshes and swamps contain relatively large quantities of grasses, trees, or shrubs; and ponds are relatively small in comparison to lakes. Geologically defined, lakes are temporary bodies of water. Check date values in: |access-date= (help)
15. Dictionary.com definition. Retrieved 2008-06-25. a body of fresh or salt water of considerable size, surrounded by land.
16. NGDC - NOAA. Volcanic Lightning. National Geophysical Data Center - NOAA. Retrieved September 21, 2007.

17. Joe Buchdahl. Atmosphere, Climate & Environment Information Programme. Ace.mmu.ac.uk. Retrieved 2013-03-09.
18. Forthofer, Ron. It›s Time To Act On Global Warming. Boulder Daily Camera. Retrieved 2013-10-28.
19. United Nations Framework Convention on Climate ChangeRetrieved August 2008.
20. Kyoto Protocol from United Nations Framework Convention on Climate Change, Retrieved August 2008.
21. Western Climate Initiative, Retrieved on Feb 12, 2009.
22. C. W. Thornthwaite, An Approach Toward a Rational Classification of Climate, Geographical Review, 38:55-94, 1948
23. Merriam-Webster Dictionary. Weather. Retrieved on 2008-06-27.
24. Glossary of Meteorology. Hydrosphere. Retrieved on 2008-06-27.
25. Glossary of Meteorology. Troposphere. Retrieved on 2008-06-27.
26. Climate. Glossary of Meteorology. American Meteorological Society. Retrieved 2008-05-14.
27. "History of life through time". University of California Museum of Paleontology.
28. Definition of Life. California Academy of Sciences. 2006. Retrieved 2007-01-07.
29. The Concise Oxford Dictionary. English Edition 1991
30. Merriam-Webster Dictionary. Merriam-Webster Dictionary. Retrieved 2009-06-21.
31. Christopherson, Robert W. (1996). Geosystems: An Introduction to Physical Geography. Prentice Hall. ISBN 0-13-505314-5.
32. Odum, E. P. (1971). Fundamentals of Ecology (Third ed.). New York: Saunders. ISBN 0-7216-6941-7.
33. Smil, V. (2000). Cycles of Life. New York: Scientific American Library. ISBN 978-0-7167-5079-6.
34. The WILD Foundation. Wild.org. Retrieved 2013-03-09.
35. No Man's Garden by Daniel B. Botkin p155-157
36. wilderness. CollinsDictionary.com. Collins English Dictionary - Complete & Unabridged 11th Edition. Retrieved November 29, 2012.
37. Why the Amazon Rainforest is So Rich in Species : News. Earthobservatory. nasa.gov. 2005-12-05. Retrieved 2013-03-09.
38. Why The Amazon Rainforest Is So Rich In Species. Sciencedaily.com. 2005-12-05. Retrieved 2013-03-09.
39. Escolero, O., Kralisch, S., Martínez, S.E., Perevochtchikova, M., (2016). Diagnóstico y análisis de los factores que influyen en la vulnerabilidad de las fuentes de abastecimiento de agua potable a la Ciudad de México, México (PDF). Boletín de la Sociedad Geológica Mexicana (in Spanish). **68** *(3): 409–427.*
40. Jamieson, Dale. (2007). The Heart of Environmentalism. In R. Sandler &

P. C. Pezzullo. Environmental Justice and Environmentalism. (pp. 85-101). Massachusetts Institute of Technology Press.

41. Davis, T. (2000). Sustaining the Forest, the People, and the Spirit. (pp. 1-24). State University of New York.

CHAPTER 11

Regime Shift

Chapter Contents

We define regime shifts as large, persistent changes in the structure and function of social-ecological systems, with substantive impacts on the suite of ecosystem services provided by these systems. Better understanding of regime shifts is important as they may have substantial impacts on human economies, societies and well-being, and are often difficult to anticipate and costly to reverse.

INTRODUCTION

In ecology, **regime shifts** are large, abrupt, persistent changes in the structure and function of a system. A regime is a characteristic behaviour of a system which is maintained by mutually reinforced processes or feedbacks. Regimes are considered persistent relative to the time period over which the shift occurs. The change of regimes, or the shift, usually occurs when a smooth change in an internal process (feedback) or a single disturbance (external shocks) triggers a completely different system behavior. Although such non-linear changes have been widely studied in different disciplines ranging from atoms to climate dynamics, regime shifts have gained importance in ecology because they can substantially affect the flow of ecosystem services that societies rely upon, such as provision of food, clean water or climate regulation. Moreover, regime shift occurrence is expected to increase as human influence on the planet increases – the Anthropocene – including current trends on human induced climate change and biodiversity loss.

HISTORY OF THE CONCEPT

Scholars have been interested in systems exhibiting non-linear change for a long time. Since the early 20th century, mathematicians have developed a body of concepts and theory for the study of such phenomena based on the study of non-linear system dynamics. This research led to the development of concepts such as catastrophe theory; a branch of bifurcation theory in dynamical systems.

In ecology the idea of systems with multiple regimes, domains of attraction called alternative stable states, only arose in the late ‹60s based upon the first reflections on the meaning of stability in ecosystems by Richard Lewontin and Crawford "Buzz" Holling. The first work on regime shifts in ecosystems was done in a diversity of ecosystems and included important work by Noy-Meir (1975) in grazing systems; May (1977) in grazing systems, harvesting systems, insect pests and host-parasitoid systems; Jones and Walters (1976) with fisheries systems; and Ludwig et al. (1978) with insect outbreaks.

These early efforts to understand regime shifts were criticized for the difficulty of demonstrating bi-stability, their reliance on simulation models, and lack of high quality long-term data. However, by the 1990s more substantial evidence of regime shifts was collected for kelp forest, coral reefs, drylands and shallow lakes. This work led to revitalization of research on ecological reorganization and the conceptual clarification that resulted in the regime shift conceptual framework in the early 2000s.

Outside of ecology, similar concepts of non-linear change have been developed in other academic disciplines. One example is historical institutionalism in political science, sociology and economics, where concepts like path dependency and critical junctures are used to explain phenomena where the output of a system is determined by its history, or the initial conditions, and where its domains of attraction are reinforced by feedbacks. Concept such as international institutional regimes, socio-technical transitions and increasing returns have an epistemological basis similar to regime shifts, and utilize similar mathematical models.

CURRENT APPLICATIONS OF THE REGIME SHIFT CONCEPT

During the last decades, research on regime shift has grown exponentially. Academic papers reported by ISI Web of Knowledge rose from less than 5 per year prior to 1990 to more than 300 per year from 2007 to 2011. However, the application of regime shift related concepts is still contested.

Although there is not agreement on one definition, the slight differences among definitions reside on the meaning of stability – the measure of what a regime is – and the meaning of abruptness. Both depend on the definition of the system under study, thus it is relative. At the end it is a matter of scale. Mass extinctions are regime shifts on the geological time scale, while financial crises or pest outbreaks are regime shifts that require a totally different parameter setting.

In order to apply the concept to a particular problem, one has to conceptually limit its range of dynamics by fixing analytical categories such as time and space scales, range of variations and exogenous / endogenous processes. For example, while for oceanographers a regime must last for at least decades and should include climate variability as a driver, for marine biologists regimes of only five years are acceptable and could be induced by only population dynamics. A non-exhaustive range of current definitions of regime shifts in recent scientific literature from ecology and allied fields is collected in Table 1.

Table 1. Definitions of regime shifts and modifications used to apply the concept to particular research questions from scientific literature published between 2004 and 2009.

Source	**Definition**	**Modification**
Collie *et al.* 2004	"Three different types of regime shift (smooth, abrupt and discontinuous) are identified on the basis of different patterns in the relationship between the response of an ecosystem variable (usually biotic) and some external forcing or condition (control variable). The smooth regime shift is represent by a quasi-linear relationship between the response and control variables. The abrupt regime shift exhibits a nonlinear relationship between the response and control variables, and the discontinuous regime shift is characterized by the trajectory of the response variable differing when the forcing variable increases compared to when it decreases (i.e., the occurrence of alternative "stable" states)"	""Regime shifts" are considered here to be low-frequency, high-amplitude changes in oceanic conditions that may be especially pronounced in biological variables and propagate through several trophic levels"
Bakun 2004 (in Collie et al. 2004)		"persistent radical shift in typical levels of abundance or productivity of multiple important components of marine biological community structure, occurring at multiple trophic levels and on a geographical scale that is at least regional in extent"
Walker & Meyers, 2004	"A regime shift involving alternate stable states occurs when a threshold level of a controlling variable in a system is passed, such that the nature and extend of feedbacks change, resulting in a change of direction (the trajectory) of the system itself. A shift occurs when internal processes of the system (…) have changed and the state of the system […] begins to change in a different direction, toward a different attractor."	
Andersen et al. 2009		"ecological regime shifts can be defined as abrupt changes on several trophic levels leading to rapid ecosystem reconfiguration between alternative states"

Cumming & Norberg, 2008	"the ability of a system to internally switch between different self reinforcing processes that dominate how the system functions"	
Brock, Carpenter and Scheffer 2008 (Chap 6 in Cumming and Norberg)	"Regime shifts, substantial reorganizations of complex systems with prolonged consequences [...] In environmental policy regime shifts raise the prospect that incremental stresses may evoke large, unexpected changes in ecosystem services and human livelihoods"	
Biggs *et al.* 2009	"Ecological regime shifts are large, sudden changes in ecosystems that last of substantial periods of time [...] Regime shifts entail changes in the internal dynamics and feedbacks of an ecosystem that often prevent it from returning to a previous regime, even when the driver that precipitated the shift is reduced or removed [...] Regime shifts typically result from a combination of gradual changes in an underlying driving variable (or set of variables), combined with an external shock, such as a storm or fire"	"We defined a regime shift as the period over which the annual increase in the planktivore (F) population exceeded 10%. In the model, regime shifts have a typical duration of ≈15 years, reflecting plausible limits on the growth rate of F"
Norström et al. 2009	"Certain conditions may ultimately result in persistent alternative stable states (ASS), which are characterized by a different set of ecosystems processes, functions and feedback mechanisms..."	"we defined phase shifts as an extensive decreases in coral cover coinciding with substantial increases in some alternative benthic organism, due to a pulse or press disturbance, that have persisted >5yr. A minimum persistence time of 5 yr was used, as this is in accordance with the timeframe of studies describing cases of phase shifts from coral to macroalgal states..."
Scheffer (2009)	"a relatively sharp change from one regime to a contrasting one, where a regime is a dynamic 'state' of a system with its characteristics stochastic fluctuations and/or cycles"	

THEORETICAL BASIS

The theoretical basis for regime shifts has been developed from the mathematics of non-linear systems. In short, regime shifts describe dynamics characterized by the possibility that a small disturbance can produce big effects. In such situations the common notion of proportionality between inputs and outputs of a system is incorrect. Conversely, the regime shift concept also emphasizes the resilience of systems – suggesting that in some situations substantial management or human impact can have little effect on a system. Regime shifts are hard to reverse and in some cases irreversible. The regime shift concept shifts analytical attention away from linearity and predictability, towards reorganization and surprise. Thus, the regime shift concept offers a framework to explore the dynamics and causal explanations of non-linear change in nature and society.

Regime shifts are triggered either by the weakening of stabilizing internal processes – feedbacks – or by external shocks which exceed the stabilizing capacity of a system.

Systems prone to regime shifts can show three different types of change: smooth, abrupt or discontinuous, depending on the configuration of processes that define a system – in particular the interaction between a system›s fast and slow processes. Smooth change can be described by a quasi-linear relationship between fast and slow processes; abrupt change shows a non-linear relationship among fast and slow variables, while discontinuous change is characterized by the difference in the trajectory on the fast variable when the slow one increases compared to when it decreases. In other words, the point at which the system flips from one regime to another is different from the point at which the system flips back. Systems that exhibit this last type of change demonstrate hysteresis. Hysteretic systems have two important properties. First, the reversal of discontinuous change requires that a system change back past the conditions at which the change first occurred. This occurs because systemic change alters feedback processes that maintain a system in a particular regime. Second, hysteresis greatly enhances the role of history in a system, and demonstrates that the system has memory – in that its dynamics are shaped by past events.

Conditions at which a system shifts its dynamics from one set of processes to another are often called thresholds. In ecology for example, a threshold is a point at which there is an abrupt change in an ecosystem quality, property or phenomenon; or where small changes in an environmental driver produce large responses in an ecosystem. Thresholds are, however, a function of several interacting parameters, thus they change in time and space. Hence, the same system can present smooth, abrupt or discontinuous change depending on its parameters› configurations. Thresholds will be present, however, only in cases where abrupt and discontinuous change is possible.

EVIDENCE

Empirical evidence has increasingly completed model based work on regime shifts. Early work on regime shifts in ecology was developed in models for predation, grazing, fisheries and inset outbreak dynamics. Since the 1980s, further development of models has been complemented by empirical evidence for regime shifts from ecosystems including kelp forest, coral reefs, drylands and lakes.

Scholars have collected evidence for regime shifts across a wide variety of ecosystems and across a range of scales. For example, at the local scale, one of the best documented examples is bush encroachment, which is thought to follow a smooth change dynamic. Bush encroachment refers to small changes in herbivory rates that can shift drylands from grassy dominated regimes towards woody dominated savannas. Encroachment has been documented to impact ecosystem services related with cattle ranching in wet savannas in Africa and South America. At the regional scale, rainforest areas in the Amazon and East Asia are thought to be at risk of shifting towards savanna regimes given the weakening of the moisture recycling feedback driven by deforestation. The shift from forest to savanna potentially affects the provision of food, fresh water, climate regulation and support for biodiversity. On the global realm, the faster retreating of the arctic ice sheet in summer time is reinforcing climate warming through the albedo feedback, potentially affecting sea water levels and climate regulation worldwide.

Aquatic systems have been heavily studied in the search for regime shifts. Lakes work like microcosms (almost closed systems) that to some extent allow experimentation and data gathering. Eutrophication is a well-documented abrupt change from clear water to murky water regimes, which leads to toxic algae blooms and reduction of fish productivity in lakes and coastal ecosystems. Eutrophication is driven by nutrient inputs, particularly those coming from fertilizers used in agriculture. It is an example of discontinuous change with hysteresis. Once the lake has shifted to a murky water regime, a new feedback of phosphorus recycling maintains the system in the eutrophic state even if nutrient inputs are significantly reduced.

Another example widely studied in aquatic and marine systems is trophic level decline in food webs. It usually implies the shift from ecosystems dominated by high numbers of predatory fish to a regime dominated by lower trophic groups like pelagic planktivores (i.e. jellyfish). Affected food webs often have impacts on fisheries productivity, a major risk of eutrophication, hypoxia, invasion of non-native species and impacts on recreational values. Hypoxia, or the development of so-called death zones, is another regime shift in aquatic and marine-coastal environments. Hypoxia, similarly to eutrophication, is driven by nutrient inputs of anthropogenic origin but also from natural origin in the form of upwellings. In high nutrient concentrations the levels of dissolved oxygen decrease, making life impossible for the majority of aquatic organisms.Impacts on ecosystem services include collapse of fisheries and the production of toxic gases for humans.

In marine systems, two well-studied regime shifts happen in coral reefs and kelp forests. Coral reefs are three-dimensional structures which work as habitat for marine biodiversity. Hard coral-dominated reefs can shift to a regime dominated by fleshy algae; but they also have been reported to shift towards soft-corals, corallimorpharians, urchin barrens or sponge-dominated regimes. Coral reef transitions are reported to affect ecosystem services like calcium fixation, water cleansing, support for biodiversity, fisheries productivity, coastline protection and recreational services. On the other hand, kelp forests are highly productive marine ecosystems found in temperate regions of the ocean. Kelp forests are characteristically dominated by brown macroalgae and host high levels of biodiversity, providing provisioning ecosystem services for both the cosmetic industry and fisheries. Such services are substantially reduced when a kelp forest shifts towards urchin barren regimes driven mainly by discharge of nutrients from the coast and overfishing. Overfishing and overharvest of keystone predators, such as sea otters, applies top-down pressure on the system. Bottom-up pressure arises from nutrient pollution.

Soil salinization is an example of a well-known regime shift in terrestrial systems. It is driven by the removal of deep root vegetation and irrigation, which causes elevation of the soil water table and the increase of soil surface salinity. Once the system flips, ecosystem services related with food production – both crops and cattle – are significantly reduced. Dryland degradation, also known as desertification, is a well-known but controversial type of regime shift. Dryland degradation occurs when the loss of vegetation transforms an ecosystem from being vegetated to being dominated by bare soils. While this shift has been proposed to be driven by a combination of farming and cattle grazing, loss of semi-nomad traditions, extension of infrastructure, reduction of managerial flexibility and other economic factors, it is controversial because it has been difficult to determine whether there is indeed a regime shift and which drivers have caused it. For example, poverty has been proposed as a driver of dry land degradation, but studies continuously find contradictory evidence. Ecosystem services affected by dry land degradation usually include low biomass productivity, thus reducing provisioning and supporting services for agriculture and water cycling.

Polar regions have been the focus on research examining the impacts of climate warming. Regime shifts in polar regions include the melting of the Greenland ice sheet and the possible collapse of the thermohaline circulation system. While the melting of the Greenland ice sheet is driven by global warming and threatens worldwide coastlines with an increase of sea level, the collapse of the thermohaline circulation is driven by the increase of fresh water in the North Atlantic which in turn weakens the density driven water transport between the tropics and polar areas. Both regime shifts have serious implications for marine biodiversity, water cycling, security of housing and infrastructure and climate regulation amongst other ecosystem services.

DETECTING REGIME SHIFTS

Using current well-known statistical methods such as average standard deviates, principal component analysis, or artificial neural networks one can detect whether a regime shift has occurred. Such analyses require long term data series and that the threshold under study has to be crossed. Hence, the answer will depend on the quality of the data; it is event-driven and only allows one to explore past trends.

Some scholars have argued based on statistical analysis of time series that certain phenomena do not correspond to regime shifts. Nevertheless, the statistical rejection of the hypothesis that a system has multiple attractors does not imply that the null hypothesis is true. In order to do so one has to prove that the system only has one attractor. In other words, evidence that data does not exhibit multiple regimes does not rule out the possibility a system could shift to an alternative regime in the future. Moreover, in management decision making, it can be risky to assume that a system has only one regime, when plausible alternative regimes have highly negative consequences.

On the other hand, a more relevant question than "has a regime shift occurred?" is to ask "is the system prone to regime shifts". This question is important because, even if they have shown smooth change in the past, their dynamics can potentially become abrupt or discontinuous in the future depending on its parameters› configuration. Such a question has been explored separately in different disciplines for different systems, pushing methods development forward (e.g. climate driven regime shifts in the ocean or the stability of food webs) and continuing to inspire new research.

FRONTIERS OF RESEARCH

Regime shift research is occurring across multiple ecosystems and at multiple scales. New areas of research include early warnings of regime shifts and new forms of modeling.

Significant efforts have been made to identify early-warning signals of regime shifts. Systems approaching a bifurcation point show a characteristic behaviour called critical slowing down that leaves a distinguishable signal in the increase of spatial and temporal autocorrelation; variance and skewness increase while variance spectra tend to lower frequencies. Researchers have explored early warning signals in lakes, climate dynamics, food webs, dry land transitions and epilepsy attacks. It remains unclear how well such signals work for all regime shifts, and if the early warnings give time enough to take appropriate managerial corrections to avoid the shift. Additionally, early warning signals also depend on intensive good-quality data series that are rare in ecology. However, researchers have used high quality data to predict regime shifts in a lake ecosystem. Changes in spatial pattern as an indicator of regime shifts have also become a topic of research.

Another front of research is the development of new approaches to modeling. Dynamic models, Bayesian belief networks, Fisher information, and fuzzy cognitive maps have been used as a tool to explore the phase space where regime shifts are likely to happen and understand the dynamics that govern dynamic thresholds. Models are useful oversimplifications of reality, whose limits are given by the current understanding of the real system as well as the assumptions of the modeler. Therefore, a deep understanding of causal relationships and the strength of feedbacks is required to capture possible regime shift dynamics. Nevertheless, such deep understanding is available only for heavily studied systems such as shallow lakes. Methods development is required to tackle the problem of limited time series data and limited understanding of system dynamics, in such a way that allow identification of the main drivers of regime shifts as well as prioritization of managerial options.

Other emerging areas of research include the role of regime shifts in the earth system, cascading consequences among regime shifts, and regime shifts in social-ecological systems.

Systems Ecology

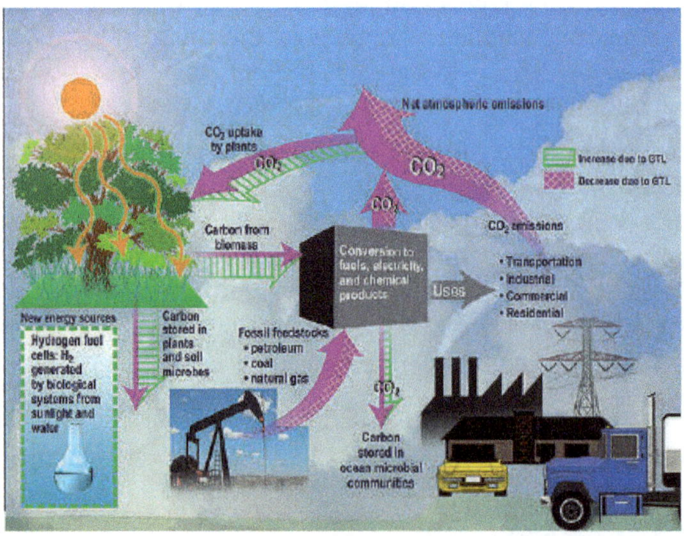

Ecological analysis of CO_2 in an ecosystem

Systems ecology is an interdisciplinary field of ecology, a subset of Earth system science, that takes a holistic approach to the study of ecological systems, especially ecosystems. Systems ecology can be seen as an application of general systems theory to ecology. Central to the systems ecology approach is the idea that an ecosystem is a complex system exhibiting emergent properties. Systems ecology focuses on interactions and transactions within and between biological and ecological systems, and is especially concerned with the way the functioning of ecosystems can be influenced by human interventions. It uses and extends concepts from thermodynamics and develops other macroscopic descriptions of complex systems.

Systems ecology seeks a holistic view of the interactions and transactions within and between biological and ecological systems. Systems ecologists realise that the function of any ecosystem can be influenced by human economics in fundamental ways. They have therefore taken an additional transdisciplinary step by including economics in the consideration of ecological-economic systems. In the words of R.L. Kitching:

- *Systems ecology can be defined as the approach to the study of ecology of organisms using the techniques and philosophy of systems analysis: that is, the methods and tools developed, largely in engineering, for studying, characteriszing and making predictions about complex entities, that is, systems..*
- *In any study of an ecological system, an essential early procedure is to draw a diagram of the system of interest ... diagrams indicate the system's* ***boundaries*** *by a solid line. Within these boundaries, series of components are isolated which have been chosen to represent that portion of the world in which the systems analyst is interested ... If there are no connections across the systems' boundaries with the surrounding* ***systems environments****, the systems are described as* closed. *Ecological work, however, deals almost exclusively with* open *systems.*

As a mode of scientific enquiry, a central feature of Systems Ecology is the general application of the principles of energetics to all systems at any scale. Perhaps the most notable proponent of this view was Howard T. Odum - sometimes considered the father of ecosystems ecology. In this approach the principles of energetics constitute ecosystem principles. Reasoning by formal analogy from one system to another enables the Systems Ecologist to see principles functioning in an analogous manner across system-scale boundaries. H.T. Odum commonly used the Energy Systems Language as a tool for making systems diagrams and flow charts.

The fourth of these principles, the principle of maximum power efficiency, takes central place in the analysis and synthesis of ecological systems. The fourth principle suggests that the most evolutionarily advantageous system function occurs when the environmental load matches the internal resistance of the system. The further the environmental load is from matching the internal resistance, the further the system is away from its sustainable steady state. Therefore, the systems ecologist engages in a task of resistance and impedance matching in ecological engineering, just as the electronic engineer would do.

Closely Related Fields

Deep Ecology

Deep ecology is an ideology whose metaphysical underpinnings are deeply concerned with the science of ecology. The term was coined by Arne Naess, a Norwegian philosopher, Gandhian scholar, and environmental activist. He argues that the prevailing approach to environmental management is anthropocentric, and that the natural environment is not only «more complex than we imagine, it is more complex than we can imagine.» Naess formulated deep ecology in 1973 at an environmental conference in Budapest.

Joanna Macy, John Seed, and others developed Naess› thesis into a branch they called *experiential deep ecology*. Their efforts were motivated by a need they perceived for the development of an "ecological self", which views the human ego as an integrated part of a living system that encompasses the individual. They sought to transcend altruism with a deeper self-interest based on biospherical equality beyond human chauvinism.

Earth Systems Engineering and Management

Earth systems engineering and management (ESEM) is a discipline used to analyze, design, engineer and manage complexenvironmental systems. It entails a wide range of subject areas including anthropology, engineering, environmental science, ethics and philosophy. At its core, ESEM looks to «rationally design and manage coupled human-natural systems in a highly integrated and ethical fashion»

Ecological Economics

Ecological economics is a transdisciplinary field of academic research that addresses the dynamic and spatial interdependence between human economies and natural ecosystems. Ecological economics brings together and connects different disciplines, within the natural and social sciences but especially between these broad areas. As the name suggests, the field is made up of researchers with a background in economics and ecology. An important motivation for the emergence of ecological economics has been criticism on the assumptions and approaches of traditional (mainstream) environmental and resource economics.

Ecological Energetics

Ecological energetics is the quantitative study of the flow of energy through ecological systems. It aims to uncover the principles which describe the propensity of such energy flows through the trophic, or 'energy availing' levels of ecological networks. In systems ecology the principles of ecosystem energy flows or "ecosystem laws" (i.e. principles of ecological energetics) are considered formally analogous to the principles of energetics.

Ecological Humanities

Ecological humanities aims to bridge the divides between the sciences and the humanities, and between Western, Eastern and Indigenous ways of knowing nature. Like ecocentric political theory, the ecological humanities are characterised by a connectivity ontology and a commitment to two fundamental axioms relating to the need to submit to ecological laws and to see humanity as part of a larger living system.

Ecosystem Ecology

A riparian forest in the White Mountains, New Hampshire (USA)

Ecosystem ecology is the integrated study of living (biotic) and non-living (abiotic) components of ecosystems and their interactions within an ecosystem framework. This science examines how ecosystems work and relates this to their components such as chemicals, bedrock, soil, plants, and animals.

Ecosystem ecology examines physical and biological structures and examines how these ecosystem characteristics interact with each other. Ultimately, this helps us understand how to maintain high quality water and economically viable commodity production. A major focus of ecosystem ecology is on functional processes, ecological mechanisms that maintain the structure and services produced by ecosystems. These include primary productivity (production of biomass), decomposition, and trophic interactions.

Studies of ecosystem function have greatly improved human understanding of sustainable production of forage, fiber, fuel, and provision of water. Functional processes are mediated by regional-to-local level climate, disturbance, and management. Thus ecosystem ecology provides a powerful framework for identifying ecological mechanisms that interact with global environmental problems, especially global warming and degradation of surface water.

This example demonstrates several important aspects of ecosystems:

- Ecosystem boundaries are often nebulous and may fluctuate in time
- Organisms within ecosystems are dependent on ecosystem level biological and physical processes

- Adjacent ecosystems closely interact and often are interdependent for maintenance of community structure and functional processes that maintain productivity and biodiversity

These characteristics also introduce practical problems into natural resource management. Who will manage which ecosystem? Will timber cutting in the forest degrade recreational fishing in the stream? These questions are difficult for land managers to address while the boundary between ecosystems remains unclear; even though decisions in one ecosystem will affect the other. We need better understanding of the interactions and interdependencies of these ecosystems and the processes that maintain them before we can begin to address these questions.

Ecosystem ecology is an inherently interdisciplinary field of study. An individual ecosystem is composed of populations of organisms, interacting within communities, and contributing to the cycling of nutrients and the flow of energy. The ecosystem is the principal unit of study in ecosystem ecology.

Population, community, and physiological ecology provide many of the underlying biological mechanisms influencing ecosystems and the processes they maintain. Flowing of energy and cycling of matter at the ecosystem level are often examined in ecosystem ecology, but, as a whole, this science is defined more by subject matter than by scale. Ecosystem ecology approaches organisms and abiotic pools of energy and nutrients as an integrated system which distinguishes it from associated sciences such as biogeochemistry.

Biogeochemistry and hydrology focus on several fundamental ecosystem processes such as biologically mediated chemical cycling of nutrients and physical-biological cycling of water. Ecosystem ecology forms the mechanistic basis for regional or global processes encompassed by landscape-to-regional hydrology, global biogeochemistry, and earth system science.

Ecosystem ecology is philosophically and historically rooted in terrestrial ecology. The ecosystem concept has evolved rapidly during the last 100 years with important ideas developed by Frederic Clements, a botanist who argued for specific definitions of ecosystems and that physiological processes were responsible for their development and persistence. Although most of Clements ecosystem definitions have been greatly revised, initially by Henry Gleason and Arthur Tansley, and later by contemporary ecologists, the idea that physiological processes are fundamental to ecosystem structure and function remains central to ecology.

Later work by Eugene Odum and Howard T. Odum quantified flows of energy and matter at the ecosystem level, thus documenting the general ideas proposed by Clements and his contemporary Charles Elton. See Figure 3.

In this model, energy flows through the whole system were dependent on biotic and abiotic interactions of each individual component (species, inorganic pools of nutrients, etc.). Later work demonstrated that these interactions and flows applied to nutrient cycles, changed over the course of succession, and held powerful controls over ecosystem productivity. Transfers of energy and nutrients are innate to ecological systems regardless of whether they are aquatic or terrestrial. Thus, ecosystem ecology has emerged from important biological studies of plants, animals, terrestrial, aquatic, and marine ecosystems.

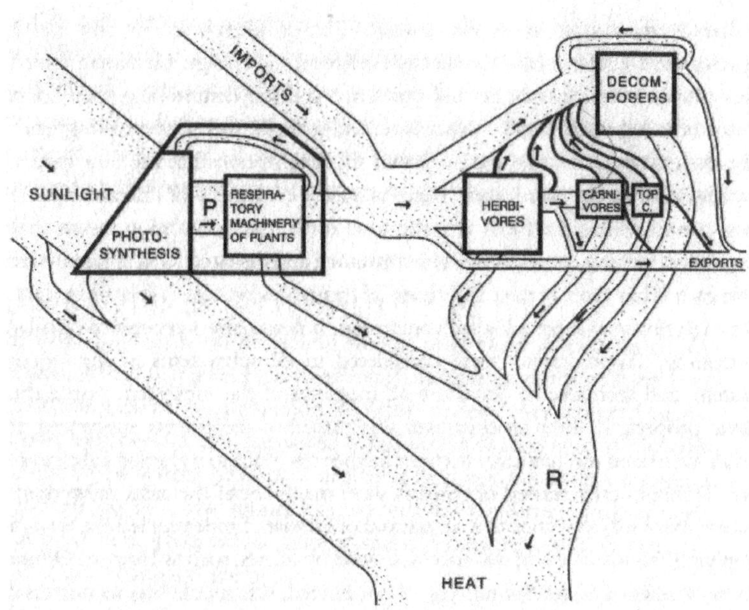

Figure 3: Energy and matter flows through an ecosystem, adapted from the Silver Springs model. H are herbivores, C are carnivores, TC are top carnivores, and D are decomposers. Squares represent biotic pools and ovals are fluxes or energy or nutrients from the system.

Ecosystem Services

Ecosystem services are ecologically mediated functional processes essential to sustaining healthy human societies. Water provision and filtration, production of biomass in forestry, agriculture, and fisheries, and removal of greenhouse gases such as carbon dioxide (CO_2) from the atmosphere are examples of ecosystem services essential to public health and economic opportunity. Nutrient cycling is a process fundamental to agricultural and forest production.

However, like most ecosystem processes, nutrient cycling is not an ecosystem characteristic which can be "dialed" to the most desirable level. Maximizing production in degraded systems is an overly simplistic solution to the complex problems of hunger and economic security. For instance, intensive fertilizer use in the midwestern United States has resulted in degraded fisheries in the Gulf of Mexico. Regrettably, a "Green Revolution" of intensive chemical fertilization has been recommended for agriculture in developed and developing countries. These strategies risk alteration of ecosystem processes that may be difficult to restore, especially when applied at broad scales without adequate assessment of impacts. Ecosystem processes may take many years to recover from significant disturbance.

For instance, large-scale forest clearance in the northeastern United States during the 18th and 19th centuries has altered soil texture, dominant vegetation, and nutrient cycling in ways that impact forest productivity in the present day. An appreciation of the importance of ecosystem function in maintenance of productivity, whether in agriculture or forestry, is needed in conjunction with plans for restoration of essential processes. Improved knowledge of ecosystem function will help to achieve long-term sustainability and stability in the poorest parts of the world.

Operation

Biomass productivity is one of the most apparent and economically important ecosystem functions. Biomass accumulation begins at the cellular level via photosynthesis. Photosynthesis requires water and consequently global patterns of annual biomass production are correlated with annual precipitation. Amounts of productivity are also dependent on the overall capacity of plants to capture sunlight which is directly correlated with plant leaf area and N content.

Net primary productivity (NPP) is the primary measure of biomass accumulation within an ecosystem. Net primary productivity can be calculated by a simple formula where the total amount of productivity is adjusted for total productivity losses through maintenance of biological processes:

$$NPP = GPP - R_{producer}$$

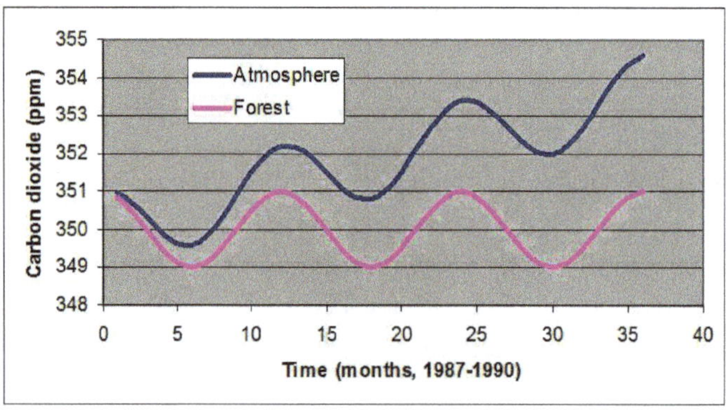

Figure : Seasonal and annual changes in ambient carbon dioxide (CO_2) concentration at Mauna Loa Hawaii (Atmosphere) and above the canopy of a deciduous forest in Massachusetts (Forest). Data show clear seasonal trends associated with periods of high and low NPP and an overall annual increase of atmospheric CO_2. Data approximates of those reported by Keeling and Whorfand Barford.

Where GPP is gross primary productivity and $R_{producer}$ is photosynthate (Carbon) lost via cellular respiration.

NPP is difficult to measure but a new technique known as eddy co-variance has shed light on how natural ecosystems influence the atmosphere. Figure 4 shows seasonal and annual changes in CO_2 concentration measured at Mauna Loa, Hawaii from 1987 to 1990. CO_2 concentration steadily increased, but within-year variation has been greater than the annual increase since measurements began in 1957.

These variations were thought to be due to seasonal uptake of CO_2 during summer months. A newly developed technique for assessing ecosystem NPP has confirmed seasonal variation are driven by seasonal changes in CO_2 uptake by vegetation. This has led many scientists and policy makers to speculate that ecosystems can be managed to ameliorate problems with global warming. This type of management may include reforesting or altering forest harvest schedules for many parts of the world.

Decomposition and Nutrient Cycling

Decomposition and nutrient cycling are fundamental to ecosystem biomass production. Most natural ecosystems are nitrogen (N) limited and biomass production is closely correlated with N turnover. Typically external input of nutrients is very low and efficient recycling of nutrients maintains productivity. Decomposition of plant litter accounts for the majority of nutrients recycled through ecosystems (Figure 3). Rates of plant litter decomposition are highly dependent on litter quality; high concentration of phenolic compounds, especially lignin, in plant litter has a retarding effect on litter decomposition. More complex C compounds are decomposed more slowly and may take many years to completely breakdown. Decomposition is typically described with exponential decay and has been related to the mineral concentrations, especially manganese, in the leaf litter.

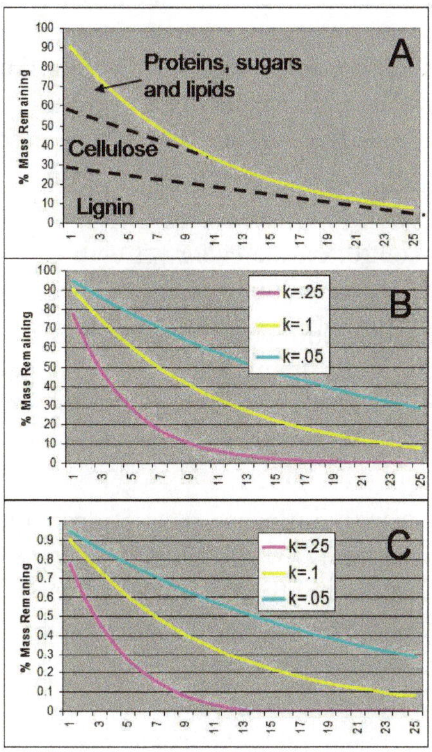

Figure 5: Dynamics of decomposing plant litter (A) described with an exponential model (B) and a combined exponential-linear model (C).

Globally, rates of decomposition are mediated by litter quality and climate.Ecosystems dominated by plants with low-lignin concentration often have rapid rates of decomposition and nutrient cycling (Chapin et al. 1982). Simple carbon (C) containing compounds are preferentially metabolized by decomposer microorganisms which results in rapid initial rates of decomposition, see Figure 5A, models that depend on constant rates of decay; so called "k" values, see Figure 5B. In addition to litter quality and climate, the activity of soil fauna is very important

However, these models do not reflect simultaneous linear and non-linear decay processes which likely occur during decomposition. For instance, proteins, sugars and lipids decompose exponentially, but lignin decays at a more linear rateThus, litter decay is inaccurately predicted by simplistic models.

A simple alternative model presented in Figure 5C shows significantly more rapid decomposition that the standard model of figure 4B. Better understanding of decomposition models is an important research area of ecosystem ecology because this process is closely tied to nutrient supply and the overall capacity of ecosystems to sequester CO_2 from the atmosphere.

Trophic Dynamics

Trophic dynamics refers to process of energy and nutrient transfer between organisms. Trophic dynamics is an important part of the structure and function of ecosystems. Figure 3 shows energy transferred for an ecosystem at Silver Springs, Florida. Energy gained by primary producers (plants, P) is consumed by herbivores (H), which are consumed by carnivores (C), which are themselves consumed by "top- carnivores"(TC).

One of the most obvious patterns in Figure 3 is that as one moves up to higher trophic levels (i.e. from plants to top-carnivores) the total amount of energy decreases. Plants exert a "bottom-up" control on the energy structure of ecosystems by determining the total amount of energy that enters the system.

However, predators can also influence the structure of lower trophic levels from the top-down. These influences can dramatically shift dominant species in terrestrial and marine systems The interplay and relative strength of top-down vs. bottom-up controls on ecosystem structure and function is an important area of research in the greater field of ecology.

Trophic dynamics can strongly influence rates of decomposition and nutrient cycling in time and in space. For example, herbivory can increase litter decomposition and nutrient cycling via direct changes in litter quality and altered dominant vegetation. Insect herbivory has been shown to increase rates of decomposition and nutrient turnover due to changes in litter quality and increased frass inputs.

However, insect outbreak does not always increase nutrient cycling. Stadler showed that C rich honeydew produced during aphid outbreak can result in increased N immobilization by soil microbes thus slowing down nutrient cycling and potentially limiting biomass production. North atlantic marine ecosystems have been greatly altered by overfishing of cod. Cod stocks crashed in the 1990s which resulted in increases in their prey such as shrimp and snow crab Human intervention in ecosystems has resulted in dramatic changes to ecosystem structure and function. These changes are occurring rapidly and have unknown consequences for economic security and human well-being.

Applications and Importance

Lessons from two Central American cities

The biosphere has been greatly altered by the demands of human societies. Ecosystem ecology plays an important role in understanding and adapting to the most pressing current environmental problems. Restoration ecology and ecosystem management are closely associated with ecosystem ecology. Restoring highly degraded resources depends on integration of functional mechanisms of ecosystems.

Without these functions intact, economic value of ecosystems is greatly reduced and potentially dangerous conditions may develop in the field. For example, areas within the mountainous western highlands of Guatemala are more susceptible to catastrophic landslides and crippling seasonal water shortages due to loss of forest resources. In contrast, cities such as Totonicapán that have preserved forests through strong social institutions have greater local economic stability and overall greater human well-being.

This situation is striking considering that these areas are close to each other, the majority of inhabitants are of Mayan descent, and the topography and overall resources are similar. This is a case of two groups of people managing resources in fundamentally different ways. Ecosystem ecology provides the basic science needed to avoid degradation and to restore ecosystem processes that provide for basic human needs.

Industrial Ecology

Industrial ecology is the study of industrial processes as linear (open loop) systems, in which resource and capital investments move through the system to become waste, to a closed loop system where wastes become inputs for new processes.

REFERENCES

1. Biggs, R., et al. (2009) Turning back from the brink: Detecting an impending regime shift in time to avert it. *P Natl Acad Sci Usa* 106, 826–831
2. Scheffer, M., et al. (2001) Catastrophic shifts in ecosystems. *Nature* 413, 591–596
3. Scheffer, M., and Carpenter, S. (2003) Catastrophic regime shifts in ecosystems: linking theory to observation. *Trends Ecol. Evol.* 18, 648–656
4. Folke, C., et al. (2004) Regime Shifts, Resilience, and Biodiversity in Ecosystem Management. *Annu. Rev. Ecol. Evol. Syst.* 35, 557–581
5. Beisner, B., et al. (2003) Alternative stable states in ecology. *Front. Ecol. Environ.* 1, 376–382

6. Feudel, U. (2008) Complex dynamics in multistable systems. *Int J Bifurcat Chaos* 18, 1607–1626

7. Biggs, R., et al. (2009) Turning back from the brink: Detecting an impending regime shift in time to avert it. *P Natl Acad Sci Usa* 106, 826–831

8. Millennium Ecosystem Assessment (2005) Ecosystems and human well-being: biodiversity synthesis. 87

9. Steffen, W., et al. (2007) The Anthropocene: Are humans now overwhelming the great forces of nature. *Ambio* 36, 614–621

10. Rockström, J., et al. (2009) A safe operating space for humanity. *Nature* 461, 472–475

11. Lewontin, R. (1969) Meaning of Stability. *Brookhaven Sym Biol*, 13

12. Holling, C.S. (1973) Resilience and stability of ecological systems. *Annual Review of Ecology and Systematics* 4, 1–23

13. Noymeir, I. (1975) Stability of Grazing Systems - Application of Predator-Prey Graphs. *Journal of Ecology* 63, 459–481

14. May, R.M. (1977) Thresholds and Breakpoints in Ecosystems with a Multiplicity of Stable States. *Nature* 269, 471–477

15. Jones, D.D., and Walters, C.J. (1976) Catastrophe Theory and Fisheries Regulation. *Journal of the Fisheries Research Board of Canada* 33, 2829–2833

16. Ludwig, D., et al. (1978) Qualitative-Analysis of Insect Outbreak Systems - Spruce Budworm and Forest. *J. Anim. Ecol.* 47, 315–332

17. Collie, J., et al. (2004) Regime shifts: can ecological theory illuminate the mechanisms? *Prog. Oceanogr.* 60, 281–302

18. Scheffer, M., et al. (2001) Catastrophic shifts in ecosystems. *Nature* 413, 591–596

19. Scheffer, M., and Carpenter, S. (2003) Catastrophic regime shifts in ecosystems: linking theory to observation. *Trends Ecol. Evol.* 18, 648–656

20. Folke, C., et al. (2004) Regime Shifts, Resilience, and Biodiversity in Ecosystem Management. *Annu. Rev. Ecol. Evol. Syst.* 35, 557–581

21. Beisner, B., et al. (2003) Alternative stable states in ecology. *Front. Ecol. Environ.* 1, 376–382

22. Collie, J., et al. (2004) Regime shifts: can ecological theory illuminate the mechanisms? *Prog. Oceanogr.* 60, 281–302

23. Norström, A., et al. (2009) Alternative states on coral reefs: beyond coral–macroalgal phase shifts. *Mar. Ecol. Prog. Ser.* 376, 295–306

24. Collie, J., et al. (2004) Regime shifts: can ecological theory illuminate the mechanisms? *Prog. Oceanogr.* 60, 281–302

25. Walker, B., and Meyers, J. (2004) Thresholds in ecological and socialecological systems: a developing database. *Ecol. Soc.* 9, 3

26. Andersen, T., et al. (2009) Ecological thresholds and regime shifts: approaches to identification. *Trends Ecol. Evol.* 24, 49–57

27. Norberg, J., and Cumming, G.S. (2008) *Complexity theory for a sustainable future*. Columbia University Press

28. Biggs, R., et al. (2009) Turning back from the brink: Detecting an impending regime shift in time to avert it. *P Natl Acad Sci Usa* 106, 826–831

29. Norström, A., et al. (2009) Alternative states on coral reefs: beyond coral–macroalgal phase shifts. *Mar. Ecol. Prog. Ser.* 376, 295–306

30. Scheffer, M. (2009) *Critical transitions in nature and society*. Princeton University Press

31. Scheffer, M., and Carpenter, S. (2003) Catastrophic regime shifts in ecosystems: linking theory to observation. *Trends Ecol. Evol.* 18, 648–656

32. Collie, J., et al. (2004) Regime shifts: can ecological theory illuminate the mechanisms? *Prog. Oceanogr.* 60, 281–302

33. Scheffer, M., et al. (2001) Catastrophic shifts in ecosystems. *Nature* 413, 591–596

34. Mayer, A., and Rietkerk, M. (2004) The dynamic regime concept for ecosystem management and restoration. *BioScience* 54, 1013–1020

35. Groffman, P., et al. (2006) Ecological thresholds: The key to successful environmental management or an important concept with no practical application? *Ecosystems* 9, 1–13

36. Folke, C., et al. (2004) Regime Shifts, Resilience, and Biodiversity in Ecosystem Management. *Annu. Rev. Ecol. Evol. Syst.* 35, 557–581

37. Roques, K., et al. (2001) Dynamics of shrub encroachment in an African savanna: relative influences of fire, herbivory, rainfall and density dependence. *J Appl Ecol* 38, 268–280

38. Anderies, J., et al. (2002) Grazing management, resilience, and the dynamics of a fire-driven rangeland system. *Ecosystems* 5, 23–44

39. Wiegand, K., et al. (2006) A patch-dynamics approach to savanna dynamics and woody plant encroachment - Insights from an arid savanna. *Perspect Plant Ecol* 7, 229–242

40. Bonan, G. (2008) Forests and climate change: Forcings, feedbacks, and the climate benefits of forests. *Science* 320, 1444–1449

41. Dekker, S.C., et al. (2010) Biogeophysical feedbacks trigger shifts in the modelled vegetation-atmosphere system at multiple scales. *Biogeosciences* 7, 1237–1245

42. Dekker, S.C., et al. (2007) Coupling microscale vegetation-soil water and macroscale vegetation-precipitation feedbacks in semiarid ecosystems. *Global Change Biol* 13, 671–678

43. Rietkerk, M., et al. (2004) Self-organized patchiness and catastrophic shifts in ecosystems. *Science* 305, 1926–1929

44. Da Silva, R., et al. (2008) Regional impacts of future land-cover changes on the amazon basin wet-season climate. *J Climate* 21, 1153–1170

45. Oyama, M., and Nobre, C. (2003) A new climate-vegetation equilibrium state for tropical South America. *Geophys Res Lett* 30, 2199

46. Holling, C.S. (1973) Resilience and stability of ecological systems. *Annual Review of Ecology and Systematics* 4, 1–23

47. Carpenter, S., and Kinne, O. (2003) Regime shifts in lake ecosystems: pattern and variation. Ecology Institute

48. Scheffer, M., and van Nes, E. (2004) Mechanisms for marine regime shifts: Can we use lakes as microcosms for oceans? *Prog. Oceanogr.* 60, 303–319

49. Carpenter, S., and Kinne, O. (2003) Regime shifts in lake ecosystems: pattern and variation. Ecology Institute

50. Carpenter, S.R., and Lathrop, R.C. (2008) Probabilistic estimate of a threshold for eutrophication. *Ecosystems* 11, 601–613

51. Carpenter, S.R., et al. (1999) Management of eutrophication for lakes subject to potentially irreversible change. *Ecol. Appl.* 9, 751–771

52. Daskalov, G.M., et al. (2007) Trophic cascades triggered by overfishing reveal possible mechanisms of ecosystem regime shifts. *P Natl Acad Sci Usa* 104, 10518–10523

53. Frank, K.T., et al. (2005) Trophic cascades in a formerly cod-dominated ecosystem. *Science* 308, 1621–1623

54. Jackson, J., et al. (2001) Historical overfishing and the recent collapse of coastal ecosystems. *Science* 293, 629–638

55. Pace, M., et al. (1999) Trophic cascades revealed in diverse ecosystems. *Trends Ecol. Evol.* 14, 483–488

56. Pauly, D., and Palomares, M. (2005) Fishing down marine food web: It is far more pervasive than we thought. *Bull. Mar. Sci.* 76, 197–211

57. Diaz, R.J., and Rosenberg, R. (2008) Spreading Dead Zones and Consequences for Marine Ecosystems. *Science* 321, 926–929

58. Hoegh-Guldberg, O., et al. (2007) Coral reefs under rapid climate change and ocean acidification. *Science* 318, 1737–1742

59. Knowlton, N. (2004) Multiple "stable" states and the conservation of marine ecosystems. *Prog. Oceanogr.* 60, 387–396

60. Knowlton, N. (1992) Thresholds and Multiple Stable States in Coral-Reef Community Dynamics. *Am Zool* 32, 674–682

61. Hughes, T., et al. (2010) Rising to the challenge of sustaining coral reef resilience. *Trends Ecol. Evol.*

62. Bellwood, D., et al. (2004) Confronting the coral reef crisis. *Nature* 429, 827–833

63. Nyström, M., and Folke, C. (2001) Spatial resilience of coral reefs. *Ecosystems* 4, 406–417

64. Norström, A., et al. (2009) Alternative states on coral reefs: beyond coral–macroalgal phase shifts. *Mar. Ecol. Prog. Ser.* 376, 295–306

65. Hoegh-Guldberg, O., et al. (2007) Coral reefs under rapid climate change and ocean acidification. *Science* 318, 1737–1742

66. Moberg, F., and Folke, C. (1999) Ecological goods and services of coral reef ecosystems. *Ecological Economics* 29, 215–233

67. Gorman, D., et al. (2009) Land-to-sea connectivity: linking human-derived terrestrial subsidies to subtidal habitat change on open rocky coasts. *Ecol. Appl.* 19, 1114–1126

68. Lauzon-Guay, J.-S., et al. (2009) Modelling phase shifts in a rocky subtidal ecosystem. *Mar Ecol-Prog Ser* 375, 25–39

69. Ling, S., et al. (2009) Overfishing reduces resilience of kelp beds to climate-driven catastrophic phase shift. *P Natl Acad Sci Usa* 106, 22341–22345

70. Steneck, R., et al. (2004) Accelerating trophic-level dysfunction in kelp forest ecosystems of the western North Atlantic. *Ecosystems* 7, 323–332

71. Konar, B., and Estes, J. (2003) The stability of boundary regions between kelp beds and deforested areas. *Ecology* 84, 174–185

72. Steneck, R., et al. (2002) Kelp forest ecosystems: biodiversity, stability, resilience and future. *Environ. Conserv.* 29, 436–459

73. Anderies, J; Ryan, P & B. Walker. 2006. Loss of Resilience, Crisis, and Institutional Change: Lessons from an Intensive Agricultural System in Southeastern Australia. *Ecosystems* 9 (6) 865:878

74. Reynolds, J., et al. (2007) Global Desertification: Building a Science for Dryland Development. *Science* 316, 847

75. Geist, H., and Lambin, E. (2004) Dynamic causal patterns of desertification. *BioScience* 54, 817–829

76. Geist, H., and Lambin, E. (2002) Proximate causes and underlying driving forces of tropical deforestation. *BioScience* 52, 143–150

77. Lambin, E., et al. (2001) The causes of land-use and land-cover change: moving beyond the myths. *Global Environ Chang* 11, 261–269

78. Greene, C.H., et al. (2008) Arctic Climate Change and Its Impacts on the Ecology of the North Atlantic. *Ecology* 89, S24–S38

79. Hátún, H., et al. (2005) Influence of the Atlantic subpolar gyre on the thermohaline circulation. *Science* 309, 1841–1844

80. Sonderegger, D.L., et al. (2009) Using SiZer to detect thresholds in ecological data. *Front. Ecol. Environ.* 7, 190–195

81. Andersen, T., et al. (2009) Ecological thresholds and regime shifts: approaches to identification. *Trends Ecol. Evol.* 24, 49–57

82. Andersen, T., et al. (2009) Ecological thresholds and regime shifts: approaches to identification. *Trends Ecol. Evol.* 24, 49–57

83. Feng, J.F., et al. (2006) Alternative attractors in marine ecosystems: A comparative analysis of fishing effects. *Ecological Modelling* 195, 377–384

84. Overland, J., et al. (2008) North Pacific regime shifts: Definitions, issues and recent transitions. *Prog. Oceanogr.* 77, 92–102

85. Rothschild, B., and Shannon, L. (2004) Regime shifts and fishery management. *Prog. Oceanogr.* 60, 397–402

86. Hilborn, R. (2007) Reinterpreting the State of Fisheries and their Management. *Ecosystems* 10, 1362–1369

87. Scheffer, M., and Carpenter, S. (2003) Catastrophic regime shifts in ecosystems: linking theory to observation. *Trends Ecol. Evol.* 18, 648–656

88. Scheffer, M., and Carpenter, S. (2003) Catastrophic regime shifts in ecosystems: linking theory to observation. *Trends Ecol. Evol.* 18, 648–656

89. Overland, J., et al. (2008) North Pacific regime shifts: Definitions, issues and recent transitions. *Prog. Oceanogr.* 77, 92–102

90. Stouffer, D., and Bascompte, J. (2010) Understanding food-web persistence from local to global scales. *Ecol Lett* 13, 154–161

91. Bastolla, U., et al. (2009) The architecture of mutualistic networks minimizes competition and increases biodiversity. *Nature* 458, 1018–U1091

92. Biggs, R., et al. (2009) Turning back from the brink: Detecting an impending regime shift in time to avert it. *P Natl Acad Sci Usa* 106, 826–831

93. Scheffer, M., et al. (2009) Early-warning signals for critical transitions. *Nature* 461, 53–59

94. Contamin, R., and Ellison, A.M. (2009) Indicators of regime shifts in ecological systems: What do we need to know and when do we need to know it? *Ecol. Appl.* 19, 799–816

95. Dakos, V., et al. (2010) Spatial correlation as leading indicator of catastrophic shifts. *Theor Ecol* 3, 163–174

96. Dakos, V., et al. (2008) Slowing down as an early warning signal for abrupt climate change. *P Natl Acad Sci Usa* 105, 14308–14312

97. van Nes, E.H., and Scheffer, M. (2007) Slow recovery from perturbations as a generic indicator of a nearby catastrophic shift. *Am. Nat.* 169, 738–747

98. van Nes, E., and Scheffer, M. (2005) Implications of spatial heterogeneity for catastrophic regime shifts in ecosystems. *Ecology* 86, 1797–1807

99. Hastings, A., and Wysham, D.B. (2010) Regime shifts in ecological systems can occur with no warning. *Ecol Lett*, 1–9

100. Scheffer, M., et al. (2009) Early-warning signals for critical transitions. *Nature* 461, 53–59

101. Dakos, V., et al. (2008) Slowing down as an early warning signal for abrupt climate change. *P Natl Acad Sci Usa* 105, 14308–14312

102. van Nes, E.H., and Scheffer, M. (2007) Slow recovery from perturbations as a generic indicator of a nearby catastrophic shift. *Am. Nat.* 169, 738–747

103. Scheffer, M., et al. (2009) Early-warning signals for critical transitions. *Nature* 461, 53–59

104. Contamin, R., and Ellison, A.M. (2009) Indicators of regime shifts in ecological

systems: What do we need to know and when do we need to know it? *Ecol. Appl.* 19, 799–816

105. Biggs, R., et al. (2009) Turning back from the brink: Detecting an impending regime shift in time to avert it. *P Natl Acad Sci Usa* 106, 826–831

106. Carpenter, S. R. et al. Early warnings of regime shifts: a whole-ecosystem experiment. *Science* 332, 1079–1082 (2011).

107. Rietkerk, M., et al. (2004) Self-organized patchiness and catastrophic shifts in ecosystems. *Science* 305, 1926–1929

108. Eppinga, M.B., et al. (2009) Linking habitat modification to catastrophic shifts and vegetation patterns in bogs. *Plant Ecol* 200, 53–68

109. Rietkerk, M., et al. (2002) Self-organization of vegetation in arid ecosystems. *Am Nat* 160, 524–530

110. Guneralp, B., and Barlas, Y. (2003) Dynamic modelling of a shallow freshwater lake for ecological and economic sustainability. *Ecological Modelling* 167, 115–138

111. Saysel, A.K., and Barlas, Y. (2001) A dynamic model of salinization on irrigated lands. *Ecological Modelling* 139, 177–199

112. Wooldridge, S., et al. (2005) Precursors for resilience in coral communities in a warming climate: a belief network approach. *Mar Ecol-Prog Ser* 295, 157–169

113. Karunanithi, A.T., et al. (2008) Detection and Assessment of Ecosystem Regime Shifts from Fisher Information. *Ecol. Soc.* 13, 15

114. Kok, K. (2009) The potential of Fuzzy Cognitive Maps for semi-quantitative scenario development, with an example from Brazil. *Global Environmental Change* 19, 122–133

CHAPTER 12

Theoretical Ecology

Chapter Contents

- **Introduction**
- **Modelling Approaches**
- **Population Ecology**
- **Spatial Ecology**
- **Ecosystem Ecology**
- **Ecophysiology**
- **Evolutionary Ecology**
- **Other Theories**
- **History**
- **Urban Ecology**

Theoretical ecology is the use of conceptual, mathematical and computational methods to address ecological problems. It uses idealised representations of ecological systems, often parameterised with real data, to investigate issues that are often intractable to experimental or observational investigation.

INTRODUCTION

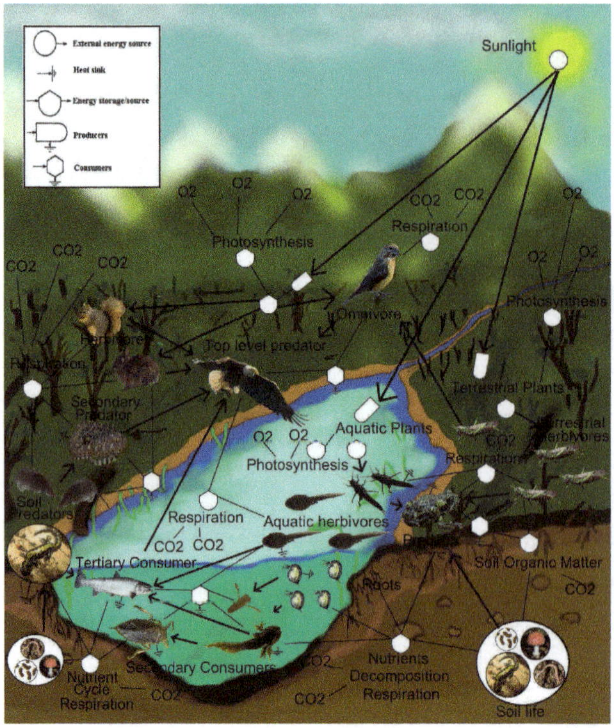

Mathematical models developed in theoretical ecology predict complex food webs are less stable than simple webs.

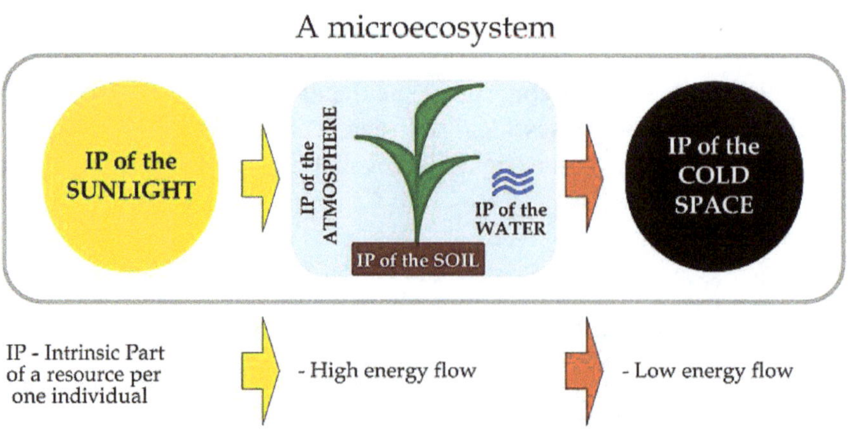

Kalmykov, L. V. & Kalmykov, V. L. A Solution to the Biodiversity Paradox by Logical Deterministic Cellular Automata. *Acta Biotheoretica* **63**, 203-221, doi:10.1007/s10441-015-9257-9 (2015).

Schematic representation of a microecosystem as a microhabitat with an individual.

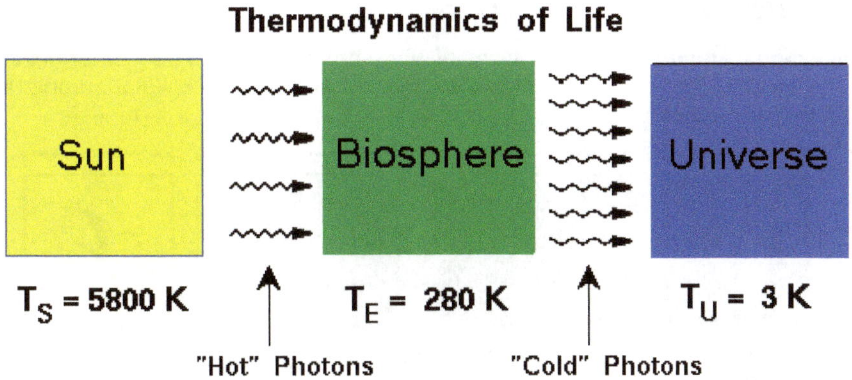

Figure 2: Life on Earth-Flow of Energy and Entropy.

Theoretical ecology is the scientific discipline devoted to the study of ecological systems using theoretical methods such as simple conceptual models, mathematical models, computational simulations, and advanced data analysis. Effective models improve understanding of the natural world by revealing how the dynamics of species populations are often based on fundamental biological conditions and processes. Further, the field aims to unify a diverse range of empirical observations by assuming that common, mechanistic processes generate observable phenomena across species and ecological environments. Based on biologically realistic assumptions, theoretical ecologists are able to uncover novel, non-intuitive insights about natural processes. Theoretical results are often verified by empirical and observational studies, revealing the power of theoretical methods in both predicting and understanding the noisy, diverse biological world.

The field is broad and includes foundations in applied mathematics, computer science, biology, statistical physics, genetics, chemistry, evolution, and conservation biology. Theoretical ecology aims to explain a diverse range of phenomena in the life sciences, such as population growth and dynamics, fisheries, competition, evolutionary theory, epidemiology, animal behavior and group dynamics, food webs, ecosystems, spatial ecology, and the effects of climate change.

Theoretical ecology has further benefited from the advent of fast computing power, allowing the analysis and visualization of large-scale computational simulations of ecological phenomena. Importantly, these modern tools provide quantitative predictions about the effects of human induced environmental change on a diverse variety of ecological phenomena, such as: species invasions, climate change, the effect of fishing and hunting on food network stability, and the global carbon cycle.

MODELLING APPROACHES

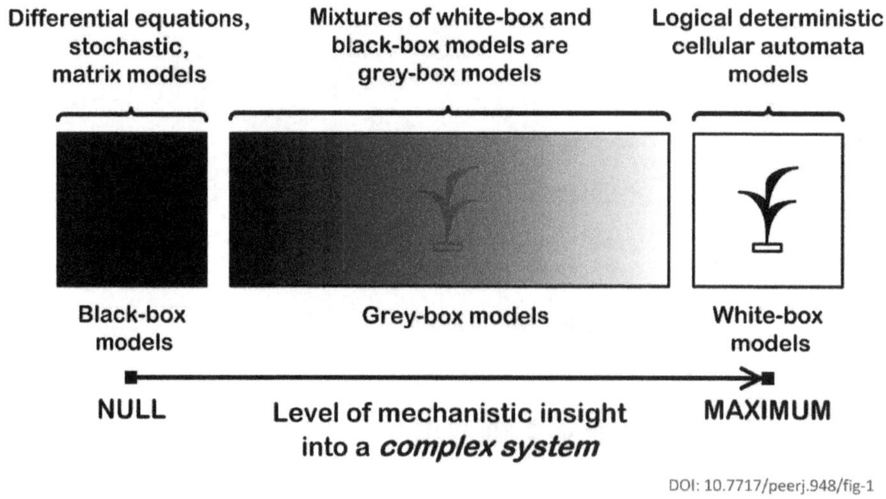

This is a schematic representation of three types of mathematical models of complex systems with the level of their mechanistic understanding.

As in most other sciences, mathematical models form the foundation of modern ecological theory.

- Phenomenological models: distill the functional and distributional shapes from observed patterns in the data, or researchers decide on functions and distribution that are flexible enough to match the patterns they or others (field or experimental ecologists) have found in the field or through experimentation.
- Mechanistic models: model the underlying processes directly, with functions and distributions that are based on theoretical reasoning about ecological processes of interest.

Ecological models can be deterministic or stochastic.

- Deterministic models always evolve in the same way from a given starting point. They represent the average, expected behavior of a system, but lack random variation. Many system dynamics models are deterministic.
- Stochastic models allow for the direct modeling of the random perturbations that underlie real world ecological systems. Markov chain models are stochastic.

Species can be modelled in continuous or discrete time.

- Continuous time is modelled using differential equations.
- Discrete time is modelled using difference equations. These model ecological processes that can be described as occurring over discrete time steps. Matrix algebra is often used to investigate the evolution of age-structured or stage-structured populations. The Leslie matrix, for example, mathematically represents the discrete time change of an age structured population.

Models are often used to describe real ecological reproduction processes of single or multiple species. These can be modelled using stochastic branching processes. Examples are the dynamics of interacting populations (predation competition and mutualism), which, depending on the species of interest, may best be modeled over either continuous or discrete time. Other examples of such models may be found in the field of mathematical epidemiology where the dynamic relationships that are to be modeled are host–pathogen interactions.

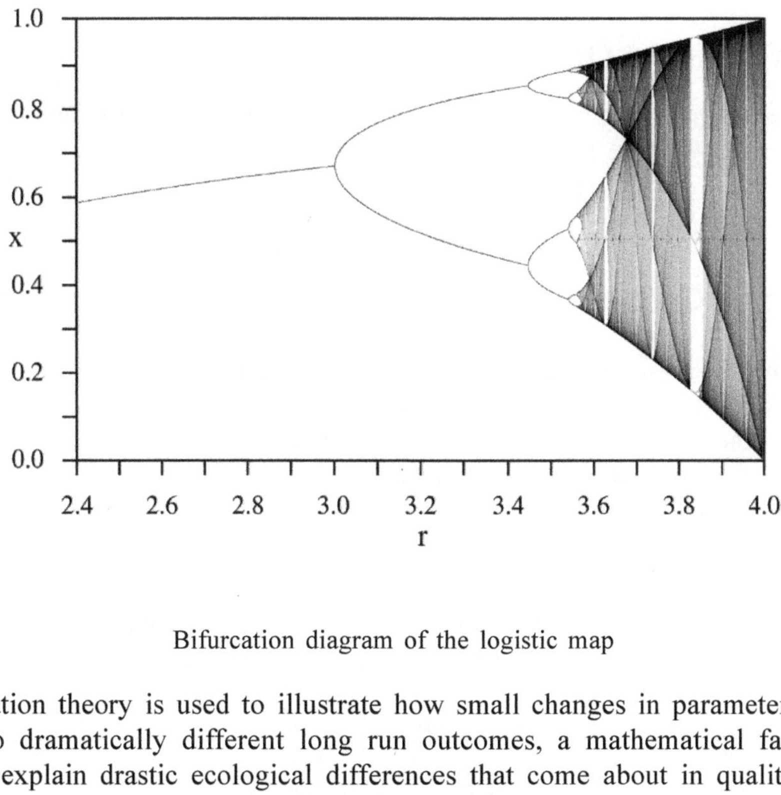

Bifurcation diagram of the logistic map

Bifurcation theory is used to illustrate how small changes in parameter values can give rise to dramatically different long run outcomes, a mathematical fact that may be used to explain drastic ecological differences that come about in qualitatively very similar systems. Logistic maps are polynomial mappings, and are often cited as providing archetypal examples of how chaotic behaviour can arise from very simple non-linear dynamical equations. The maps were popularized in a seminal 1976 paper by the theoretical ecologist Robert May. The difference equation is intended to capture the two effects of reproduction and starvation.

In 1930, R.A. Fisher published his classic *The Genetical Theory of Natural Selection*, which introduced the idea that frequency-dependent fitness brings a strategic aspect to evolution, where the payoffs to a particular organism, arising from the interplay of all of the relevant organisms, are the number of this organism' s viable offspring. In 1961, Richard Lewontin applied game theory to evolutionary biology in his *Evolution and the Theory of Games*, followed closely by John Maynard Smith, who in his seminal 1972 paper, "Game Theory and the Evolution of Fighting", defined the concept of the evolutionarily stable strategy.

Because ecological systems are typically nonlinear, they often cannot be solved analytically and in order to obtain sensible results, nonlinear, stochastic and computational techniques must be used. One class of computational models that is becoming increasingly popular are the agent-based models. These models can simulate the actions and interactions of multiple, heterogeneous, organisms where more traditional, analytical techniques are inadequate. Applied theoretical ecology yields results which are used in the real world. For example, optimal harvesting theory draws on optimization techniques developed in economics, computer science and operations research, and is widely used in fisheries.

POPULATION ECOLOGY

Population ecology is a sub-field of ecology that deals with the dynamics of species populations and how these populations interact with the environment. It is the study of how the population sizes of species living together in groups change over time and space, and was one of the first aspects of ecology to be studied and modelled mathematically.

Exponential Growth

The most basic way of modeling population dynamics is to assume that the rate of growth of a population depends only upon the population size at that time and the per capita growth rate of the organism. In other words, if the number of individuals in a population at a time t, is N(t), then the rate of population growth is given by:

$$\frac{dN(t)}{dt} = rN(t)$$

where r is the per capita growth rate, or the intrinsic growth rate of the organism. It can also be described as r = b-d, where b and d are the per capita time-invariant birth and death rates, respectively. This first order linear differential equation can be solved to yield the solution

$$N(t) = N(0)\, e^{rt},$$

a trajectory known as Malthusian growth, after Thomas Malthus, who first described its dynamics in 1798. A population experiencing Malthusian growth follows an exponential curve, where N(0) is the initial population size. The population grows when $r > 0$, and declines when $r < 0$. The model is most applicable in cases where a few organisms have begun a colony and are rapidly growing without any limitations or restrictions impeding their growth (e.g. bacteria inoculated in rich media).

Logistic Growth

The exponential growth model makes a number of assumptions, many of which often do not hold. For example, many factors affect the intrinsic growth rate and is often not time-invariant. A simple modification of the exponential growth is to assume that the intrinsic growth rate varies with population size. This is reasonable: the larger the population size, the fewer resources available, which can result in a lower birth rate and higher death rate. Hence, we can replace the time-invariant r with r'(t) = (b –a*N(t)) – (d + c*N(t)), where a and c are constants that modulate birth and death rates in a population dependent manner (e.g. intraspecific competition). Both a and c will depend on other environmental factors which, we can for now, assume to be constant in this approximated model. The differential equation is now:

$$\frac{dN(t)}{dt} = ((b - aN(t)) - (d - cN(t)))N(t)$$

This can be rewritten as:

$$\frac{dN(t)}{dt} = rN(t)\left(1 - \frac{N}{K}\right)$$

where r = b-d and K = (b-d)/(a+c).

The biological significance of K becomes apparent when stabilities of the equilibria of the system are considered. The constant K is the carrying capacity of the population. The equilibria of the system are N = 0 and N = K. If the system is linearized, it can be seen that N = 0 is an unstable equilibrium while K is a stable equilibrium.

Structured Population Growth

Another assumption of the exponential growth model is that all individuals within a population are identical and have the same probabilities of surviving and of reproducing. This is not a valid assumption for species with complex life histories. The exponential growth model can be modified to account for this, by tracking the number of individuals in different age classes (e.g. one-, two-, and three-year-olds) or different stage classes (juveniles, sub-adults, and adults) separately, and allowing individuals in each group to have their own survival and reproduction rates. The general form of this model is

$$\mathbf{N}_{t+1} = \mathbf{L}\mathbf{N}_t$$

where $\mathbf{N}_t$ is a vector of the number of individuals in each class at time *t* and **L** is a matrix that contains the survival probability and fecundity for each class. The matrix **L** is referred to as the Leslie matrix for age-structured models, and as the Lefkovitch matrix for stage-structured models. If parameter values in **L** are estimated from demographic data on a specific population, a structured model can then be used to predict whether this population is expected to grow or decline in the long-term, and what the expected age distribution within the population will be. This has been done for a number of species including loggerhead sea turtles and right whales.

Individual-based Population Growth

Logical deterministic individual-based cellular automata model of an ecosystem with one species. The model demonstrates a mechanism of S-shaped population growth.

Logical deterministic individual-based cellular automata model of interspecific competition for a single limited resource. A mechanism of competitive exclusion of one species by another.

Community Ecology

An ecological community is a group of trophically similar, sympatric species that actually or potentially compete in a local area for the same or similar resources. Interactions between these species form the first steps in analyzing more complex dynamics of ecosystems. These interactions shape the distribution and dynamics of species. Of these interactions, predation is one of the most widespread population activities. Taken in its most general sense, predation comprises predator–prey, host–pathogen, and host–parasitoid interactions.

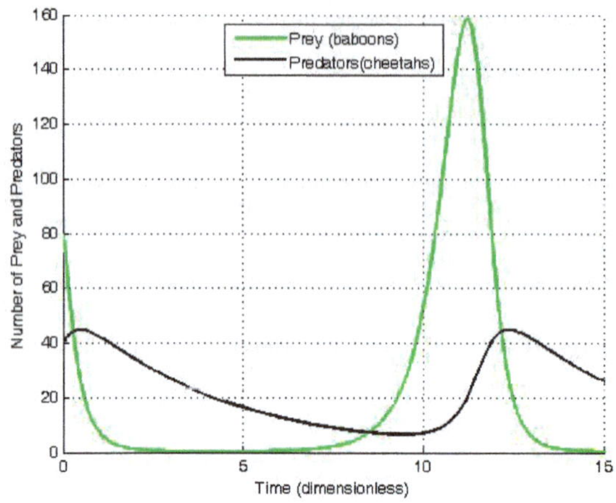

Lotka–Volterra model of cheetah–baboon interactions. Starting with 80 baboons (green) and 40 cheetahs, this graph shows how the model predicts the two species numbers will progress over time.

Predator–Prey Interaction

Predator–prey interactions exhibit natural oscillations in the populations of both predator and the prey. In 1925, the American mathematician Alfred J. Lotka developed simple equations for predator–prey interactions in his book on biomathematics. The following year, the Italian mathematician Vito Volterra, made a statistical analysis of fish catches in the Adriatic and independently developed the same equations. It is one of the earliest and most recognised ecological models, known as the Lotka-Volterra model:

$$\frac{dN(t)}{dt} = N(t)(r - \alpha P(t))$$

$$\frac{dP(t)}{dt} = P(t)(c\alpha N(t) - d)$$

where N is the prey and P is the predator population sizes, r is the rate for prey growth, taken to be exponential in the absence of any predators, α is the prey mortality rate for per-capita predation (also called 'attack rate'), c is the efficiency of conversion from prey to predator, and d is the exponential death rate for predators in the absence of any prey.

Volterra originally used the model to explain fluctuations in fish and shark populations after fishing was curtailed during the First World War. However, the equations have subsequently been applied more generally. Other examples of these models include the Lotka-Volterra model of the snowshoe hare and Canadian lynx in North America, any infectious disease modeling such as the recent outbreak of SARS and biological control of California red scale by the introduction of its parasitoid, *Aphytis melinus* .

A credible, simple alternative to the Lotka-Volterra predator–prey model and their common prey dependent generalizations is the ratio dependent or Arditi-Ginzburg model. The two are the extremes of the spectrum of predator interference models. According to the authors of the alternative view, the data show that true interactions in nature are so far from the Lotka–Volterra extreme on the interference spectrum that the model can simply be discounted as wrong. They are much closer to the ratio-dependent extreme, so if a simple model is needed one can use the Arditi–Ginzburg model as the first approximation.

Host–Pathogen Interaction

The second interaction, that of host and pathogen, differs from predator–prey interactions in that pathogens are much smaller, have much faster generation times, and require a host to reproduce. Therefore, only the host population is tracked in host–pathogen models. Compartmental models that categorize host population into groups such as susceptible, infected, and recovered (SIR) are commonly used.

Host–parasitoid Interaction

The third interaction, that of host and parasitoid, can be analyzed by the Nicholson-Bailey model, which differs from Lotka-Volterra and SIR models in that it is discrete in time. This model, like that of Lotka-Volterra, tracks both populations explicitly. Typically, in its general form, it states:

$$N_{t+1} = \lambda\, N_t\, [1 - f(N_t, P_t)]$$
$$P_{t+1} = c\, N_t\, f(N_t, p_t)$$

where $f(N_t, P_t)$ describes the probability of infection (typically, Poisson distribution), λ is the per-capita growth rate of hosts in the absence of parasitoids, and c is the conversion efficiency, as in the Lotka-Volterra model.

Competition and Mutualism

In studies of the populations of two species, the Lotka-Volterra system of equations has been extensively used to describe dynamics of behavior between two species, N_1 and N_2. Examples include relations between *D. discoiderum* and *E. coli*, as well as theoretical analysis of the behavior of the system.

$$\frac{dN_1}{dt} = \frac{r_1 N_1}{K_1}(K_1 - N_1 + \alpha_{12} N_2)$$
$$\frac{dN_2}{dt} = \frac{r_2 N_2}{K_2}(K_2 - N_2 + \alpha_{21} N_1)$$

The r coefficients give a "base" growth rate to each species, while K coefficients correspond to the carrying capacity. What can really change the dynamics of a system, however are the α terms. These describe the nature of the relationship between the two species. When α_{12} is negative, it means that N_2 has a negative effect on N_1, by competing with it, preying on it, or any number of other possibilities. When α_{12} is positive, however, it means that N_2 has a positive effect on N_1, through some kind of mutualistic interaction between the two. When both α_{12} and α_{21} are negative, the relationship is described as competitive. In this case, each species detracts from the other, potentially over competition for scarce resources. When both α_{12} and α_{21} are positive, the relationship becomes one of mutualism. In this case, each species provides a benefit to the other, such that the presence of one aids the population growth of the other.

Neutral Theory

Unified neutral theory is a hypothesis proposed by Stephen Hubbell in 2001. The hypothesis aims to explain the diversity and relative abundance of species in ecological communities, although like other neutral theories in ecology, Hubbell's hypothesis assumes that the differences between members of an ecological community of trophically similar species are neutral, or irrelevant to their success. Neutrality means that at a given trophic level in a food web, species are equivalent in birth rates, death rates, dispersal rates and speciation rates, when measured on a per-capita basis. This implies that biodiversity arises at random, as each species follows a random walk. This can be considered a null hypothesis to niche theory. The hypothesis has sparked controversy, and some authors consider it a more complex version of other null models that fit the data better.

Under unified neutral theory, complex ecological interactions are permitted among individuals of an ecological community (such as competition and cooperation), providing all individuals obey the same rules. Asymmetric phenomena such as parasitism and predation are ruled out by the terms of reference; but cooperative strategies such as swarming, and negative interaction such as competing for limited food or light are allowed, so long as all individuals behave the same way. The theory makes predictions that have implications for the management of biodiversity, especially the management of rare species. It predicts the existence of a fundamental biodiversity constant, conventionally written θ, that appears to govern species richness on a wide variety of spatial and temporal scales.

Hubbell built on earlier neutral concepts, including MacArthur & Wilson›s theory of island biogeography and Gould›s concepts of symmetry and null models.

Competition, Coexistence and Catastrophical Extinction

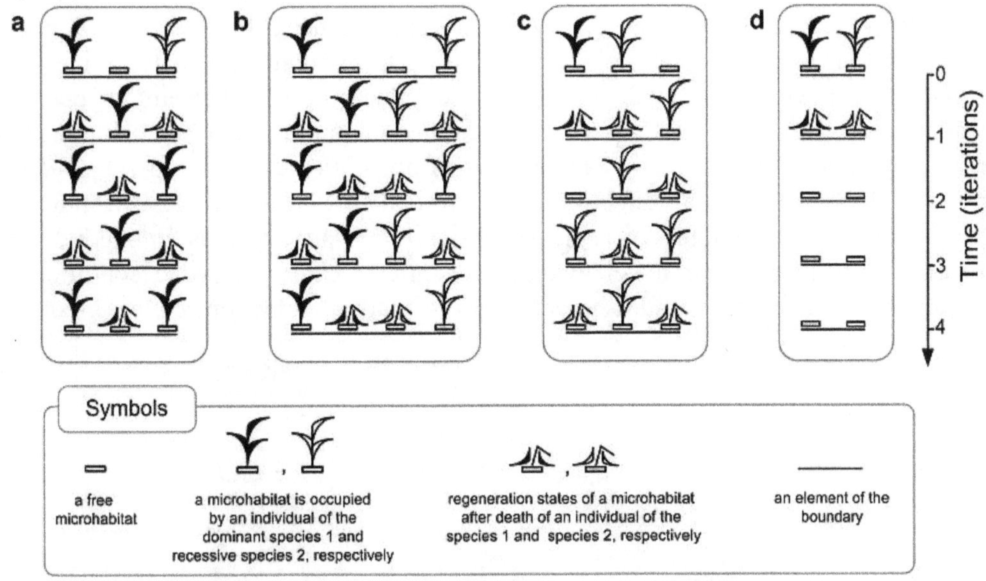

Kalmykov, L. V. & Kalmykov, V. L. A Solution to the Biodiversity Paradox by Logical Deterministic Cellular Automata. *Acta Biotheoretica* **63**, 203-221, doi:10.1007/s10441-015-9257-9 (2015).

Individual-based mechanisms of the two-species resource competition in one-dimensional habitats with boundaries. Logical deterministic cellular automata models.

SPATIAL ECOLOGY

Biogeography

Biogeography is the study of the distribution of species in space and time. It aims to reveal where organisms live, at what abundance, and why they are (or are not) found in a certain geographical area.

Biogeography is most keenly observed on islands, which has led to the development of the subdiscipline of island biogeography. These habitats are often a more manageable areas of study because they are more condensed than larger ecosystems on the mainland. In 1967, Robert MacArthur and E.O. Wilson published *The Theory of Island Biogeography*. This showed that the species richness in an area could be predicted in terms of factors such as habitat area, immigration rate and extinction rate. The theory is considered one of the fundamentals of ecological theory. The application of island biogeography theory to habitat fragments spurred the development of the fields of conservation biology and landscape ecology.

r/K-selection Theory

A population ecology concept is r/K selection theory, one of the first predictive models in ecology used to explain life-history evolution. The premise behind the r/K selection model is that natural selection pressures change according to population density. For example, when an island is first colonized, density of individuals is low. The initial increase in population size is not limited by competition, leaving an abundance of available resources for rapid population growth. These early phases of population growth experience *density-independent* forces of natural selection, which is called *r*-selection. As the population becomes more crowded, it approaches the island's carrying capacity, thus forcing individuals to compete more heavily for fewer available resources. Under crowded conditions, the population experiences density-dependent forces of natural selection, called *K*-selection.

The diversity and containment of coral reef systems make them good sites for testing niche and neutral theories.

Niche Theory

Metapopulations

Spatial analysis of ecological systems often reveals that assumptions that are valid for spatially homogenous populations – and indeed, intuitive – may no longer be valid when migratory subpopulations moving from one patch to another are considered. In a simple one-species formulation, a subpopulation may occupy a patch, move from one patch to another empty patch, or die out leaving an empty patch behind. In such a case, the proportion of occupied patches may be represented as

$$\frac{dp}{dt} = mp(1-p) - ep$$

where m is the rate of colonization, and e is the rate of extinction. In this model, if e <

m, the steady state value of p is 1 – (e/m) while in the other case, all the patches will eventually be left empty. This model may be made more complex by addition of another species in several different ways, including but not limited to game theoretic approaches, predator–prey interactions, etc. We will consider here an extension of the previous one-species system for simplicity. Let us denote the proportion of patches occupied by the first population as p_1, and that by the second as p_2. Then,

$$\frac{dp_1}{dt} = m_1 p_1(1 - p_1) - ep_1$$

$$\frac{dp_2}{dt} = m_2 p_2(1 - p_1 - p_2) - ep_2 - mp_1 p_2$$

In this case, if e is too high, p_1 and p_2 will be zero at steady state. However, when the rate of extinction is moderate, p_1 and p_2 can stably coexist. The steady state value of p_2 is given by

$$p_2^* = \frac{e}{m_1} - \frac{m_1}{m_2}$$

(p^*_1 may be inferred by symmetry). It is interesting to note that if e is zero, the dynamics of the system favor the species that is better at colonizing (i.e. has the higher m value). This leads to a very important result in theoretical ecology known as the Intermediate Disturbance Hypothesis, where the biodiversity (the number of species that coexist in the population) is maximized when the disturbance (of which e is a proxy here) is not too high or too low, but at intermediate levels.

The form of the differential equations used in this simplistic modelling approach can be modified. For example:

- Colonization may be dependent on p linearly (m*(1-p)) as opposed to the non-linear m*p*(1-p) regime described above. This mode of replication of a species is called the "rain of propagules", where there is an abundance of new individuals entering the population at every generation. In such a scenario, the steady state where the population is zero is usually unstable.
- Extinction may depend non-linearly on p (e*p*(1-p)) as opposed to the linear (e*p) regime described above. This is referred to as the "rescue effect" and it is again harder to drive a population extinct under this regime.

The model can also be extended to combinations of the four possible linear or non-linear dependencies of colonization and extinction on p are described in more detail in.

ECOSYSTEM ECOLOGY

Introducing new elements, whether biotic or abiotic, into ecosystems can be disruptive. In some cases, it leads to ecological collapse, trophic cascades and the death of many species within the ecosystem. The abstract notion of ecological health attempts to measure the robustness and recovery capacity for an ecosystem; i.e. how far the ecosystem is away from its steady state. Often, however, ecosystems rebound from a disruptive agent. The difference between collapse or rebound depends on the toxicity of the introduced element and the resiliency of the original ecosystem.

If ecosystems are governed primarily by stochastic processes, through which its subsequent state would be determined by both predictable and random actions, they may be more resilient to sudden change than each species individually. In the absence of a balance of nature, the species composition of ecosystems would undergo shifts that would depend on the nature of the change, but entire ecological collapse would probably be infrequent events. In 1997, Robert Ulanowicz used information theory tools to describe the structure of ecosystems, emphasizing mutual information (correlations) in studied systems. Drawing on this methodology and prior observations of complex ecosystems, Ulanowicz depicts approaches to determining the stress levels on ecosystems and predicting system reactions to defined types of alteration in their settings (such as increased or reduced energy flow, and eutrophication.

Ecopath is a free ecosystem modelling software suite, initially developed by NOAA, and widely used in fisheries management as a tool for modelling and visualising the complex relationships that exist in real world marine ecosystems.

Food Webs

Food webs provide a framework within which a complex network of predator–prey interactions can be organised. A food web model is a network of food chains. Each food chain starts with a primary producer or autotroph, an organism, such as a plant, which is able to manufacture its own food. Next in the chain is an organism that feeds on the primary producer, and the chain continues in this way as a string of successive predators. The organisms in each chain are grouped into trophic levels, based on how many links they are removed from the primary producers. The length of the chain, or trophic level, is a measure of the number of species encountered as energy or nutrients move from plants to top predators. Food energy flows from one organism to the next and to the next and so on, with some energy being lost at each level. At a given trophic level there may be one species or a group of species with the same predators and prey.

In 1927, Charles Elton published an influential synthesis on the use of food webs, which resulted in them becoming a central concept in ecology. In 1966, interest in food webs increased after Robert Paine›s experimental and descriptive study of intertidal shores, suggesting that food web complexity was key to maintaining species diversity and ecological stability.Many theoretical ecologists, including Sir Robert May and Stuart Pimm, were prompted by this discovery and others to examine the mathematical properties of food webs. According to their analyses, complex food webs should be less stable than simple food webs. The apparent paradox between the complexity of food webs observed in nature and the mathematical fragility of food web models is currently an area of intensive study and debate. The paradox may be due partially to conceptual differences between persistence of a food web and equilibrial stability of a food web.

Systems Ecology

Systems ecology can be seen as an application of general systems theory to ecology. It takes a holistic and interdisciplinary approach to the study of ecological systems, and particularly ecosystems. Systems ecology is especially concerned with the way the functioning of ecosystems can be influenced by human interventions. Like other fields in theoretical ecology, it uses and extends concepts from thermodynamics and develops other macroscopic descriptions of complex systems. It also takes account of the energy flows through the different trophic levels in the ecological networks. In systems ecology the principles of ecosystem energy flows are considered formally analogous to the principles of energetics. Systems ecology also considers the external influence of ecological economics, which usually is not otherwise considered in ecosystem ecology. For the most part, systems ecology is a subfield of ecosystem ecology.

ECOPHYSIOLOGY

Behavioral Ecology

Swarm Behaviour

Flocks of birds can abruptly change their direction in unison, and then, just as suddenly, make a unanimous group decision to land.

Swarm behaviour is a collective behaviour exhibited by animals of similar size which aggregate together, perhaps milling about the same spot or perhaps migrating in some direction. Swarm behaviour is commonly exhibited by insects, but it also occurs in the flocking of birds, the schooling of fish and the herd behaviour of quadrupeds. It is a complex emergent behaviour that occurs when individual agents follow simple behavioral rules.

Recently, a number of mathematical models have been discovered which explain many aspects of the emergent behaviour. Swarm algorithms follow a Lagrangian approach or an Eulerian approach. The Eulerian approach views the swarm as a field, working with the density of the swarm and deriving mean field properties. It is a hydrodynamic approach, and can be useful for modelling the overall dynamics of large swarms. However, most models work with the Lagrangian approach, which is an agent-based model following the individual agents (points or particles) that make up the swarm. Individual particle models can follow information on heading and spacing that is lost in the Eulerian approach. Examples include ant colony optimization, self-propelled particles and particle swarm optimization

EVOLUTIONARY ECOLOGY

The British biologist Alfred Russel Wallace is best known for independently proposing a theory of evolution due to natural selection that prompted Charles Darwin to publish his own theory. In his famous 1858 paper, Wallace proposed natural selection as a kind of feedback mechanism which keeps species and varieties adapted to their environment.

The action of this principle is exactly like that of the centrifugal governor of the steam engine, which checks and corrects any irregularities almost before they become evident; and in like manner no unbalanced deficiency in the animal kingdom can ever reach any conspicuous magnitude, because it would make itself felt at the very first step, by rendering existence difficult and extinction almost sure soon to follow.

The cybernetician and anthropologist Gregory Bateson observed in the 1970s that, though writing it only as an example, Wallace had "probably said the most powerful thing that'd been said in the 19th Century". Subsequently, the connection between natural selection and systems theory has become an area of active research.

OTHER THEORIES

In contrast to previous ecological theories which considered floods to be catastrophic events, the river flood pulse concept argues that the annual flood pulse is the most important aspect and the most biologically productive feature of a river›s ecosystem.

HISTORY

Theoretical ecology draws on pioneering work done by G. Evelyn Hutchinson and his students. Brothers H.T. Odum and E.P. Odum are generally recognised as the founders of modern theoretical ecology. Robert MacArthur brought theory to community ecology. Daniel Simberloff was the student of E.O. Wilson, with whom MacArthur collaborated on *The Theory of Island Biogeography*, a seminal work in the development of theoretical ecology.

Simberloff added statistical rigour to experimental ecology and was a key figure in the SLOSS debate, about whether it is preferable to protect a single large or several small reserves. This resulted in the supporters of Jared Diamond's community assembly rules defending their ideas through Neutral Model Analysis. Simberloff also played a key role in the (still ongoing) debate on the utility of corridors for connecting isolated reserves.

Stephen Hubbell and Michael Rosenzweig combined theoretical and practical elements into works that extended MacArthur and Wilson's Island Biogeography Theory - Hubbell with his Unified Neutral Theory of Biodiversity and Biogeography and Rosenzweig with his Species Diversity in Space and Time.

Theoretical and mathematical ecologists

A tentative distinction can be made between mathematical ecologists, ecologists who apply mathematics to ecological problems, and mathematicians who develop the mathematics itself that arises out of ecological problems.

Some notable theoretical ecologists can be found in these categories:

- Category:Mathematical ecologists
- Category:Theoretical biologists

URBAN ECOLOGY

Urban ecology is the scientific study of the relation of living organisms with each other and their surroundings in the context of an urban environment. The urban environment refers to environments dominated by high-density residential and commercial buildings, paved surfaces, and other urban-related factors that create a unique landscape dissimilar to most previously studied environments in the field of ecology.

Urban ecology is a recent field of study compared to ecology as a whole. The methods and studies of urban ecology are similar to and comprise a subset of ecology. The study of urban ecology carries increasing importance because, more than 50% of the world›s population today is living in urban areas. At the same time, it is estimated that within the next forty years, two-thirds of the world›s population will be living in expanding urban centers.The ecological processes in the urban environment are comparable to those outside the urban context. However, the types of urban habitats and the species that inhabit them are poorly documented. Often, explanations for phenomena examined in the urban setting as well as predicting changes because of urbanization are the center for scientific research.

Central Park represents an ecosystem fragment within a larger urban environment.

The creation of an important stream water garden in Metz's centre during the early 70s was one of the materializations of Jean-Marie Pelt's works on urban ecology.

Ecology has historically focused on "pristine" natural environments, but by the 1970s many ecologists began to turn their interest towards ecological interactions taking place in, and caused by urban environments. Jean-Marie Pelt's 1977 book *The Re-Naturalized Human*, Brian Davis' 1978 publication *Urbanization and the diversity of insects*, and Sukopp et al.'s 1979 article "The soil, flora and vegetation of Berlin's wastelands" are some of the first publications to recognize the importance of urban ecology as a separate and distinct form of ecology the same way one might see landscape ecology as different from population ecology. Forman and Godron›s 1986 book *Landscape Ecology* first distinguished urban settings and landscapes from other landscapes by dividing all landscapes into five broad types. These types were divided by the intensity of human influence ranging from pristine natural environments to urban centers.

Urban ecology is recognized as a diverse and complex concept which differs in application between North America and Europe. The European concept of urban ecology examines the biota of urban areas, while the North American concept has traditionally examined the social sciences of the urban landscape, as well as the ecosystem fluxes and processes.

Urban Ecology Methods

Since urban ecology is a subfield of ecology, many of the techniques are similar to that of ecology. Ecological study techniques have been developed over centuries, but many of the techniques use for urban ecology are more recently developed. Methods used for studying urban ecology involve chemical and biochemical techniques, temperature recording, heat mapping remote sensing, and long-term ecological research sites.

Chemical and Biochemical Techniques

Chemical techniques may be used to determine pollutant concentrations and their effects. Tests can be as simple as dipping a manufactured test strip, as in the case of pH testing, or be more complex, as in the case of examining the spatial and temporal variation of heavy metal contamination due to industrial runoff. In that particular study, livers of birds from many regions of the North Sea were ground up and mercury was extracted. Additionally, mercury bound in feathers was extracted from both live birds and from museum specimens to test for mercury levels across many decades. Through these two different measurements, researchers were able to make a complex picture of the spread of mercury due to industrial runoff both spatially and temporally.

Other chemical techniques include tests for nitrates, phosphates, sulfates, etc. which are commonly associated with urban pollutants such as fertilizer and industrial byproducts. These biochemical fluxes are studied in the atmosphere (e.g. greenhouse gasses), aquatic ecosystems and soil vegetation. Broad reaching effects of these biochemical fluxes can be seen in various aspects of both the urban and surrounding rural ecosystems.

Temperature data and Heat Mapping

Temperature data can be used for various kinds of studies. An important aspect of temperature data is the ability to correlate temperature with various factors that may be affecting or occurring in the environment. Oftentimes, temperature data is collected long-term by the Office of Oceanic and Atmospheric Research (OAR), and made available to the scientific community through the National Oceanic and Atmospheric Administration (NOAA). Data can be overlaid with maps of terrain, urban features, and other spatial areas to create heat maps. These heat maps can be used to view trends and distribution over time and space.

Remote Sensing

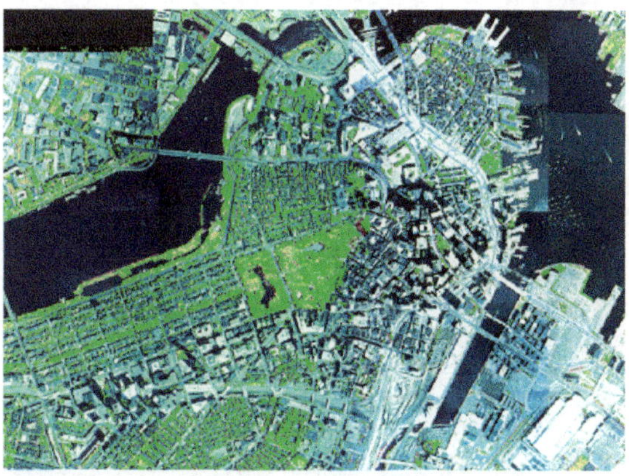

Remote sensing allows collection of data using satellites. This map shows urban tree canopy in Boston.

Remote sensing is the technique in which data is collected from distant locations through the use of satellite imaging, radar, and aerial photographs. In urban ecology, remote sensing is used to collect data about terrain, weather patterns, light, and vegetation. One application of remote sensing for urban ecology is to detect the productivity of an area by measuring the photosynthetic wavelengths of emitted light. Satellite images can also be used to detect differences in temperature and landscape diversity to detect the effects of urbanization.

LTERs and long-Term Data Sets

Long-term ecological research (LTER) sites are research sites funded by the government that have collected reliable long-term data over an extended period of time in order to identify long-term climatic or ecological trends. These sites provide long-term temporal and spatial data such as average temperature, rainfall and other ecological processes. The main purpose of LTERs for urban ecologists is the collection of vast amounts of data over long periods of time. These long-term data sets can then be analyzed to find trends relating to the effects of the urban environment on various ecological processes, such as species diversity and abundance over time. Another example is the examination of temperature trends that are accompanied with the growth of urban centers.

Urban Effects on the Environment

Humans are the driving force behind urban ecology and influence the environment in a variety of ways, such as modifying land surfaces and waterways, introducing foreign species, and altering biogeochemical cycles. Some of these effects are more apparent, such as the reversal of the Chicago River to accommodate the growing pollution levels and trade on the river.Other effects can be more gradual such as the change in global climate due to urbanization.

Modification of Land and Waterways

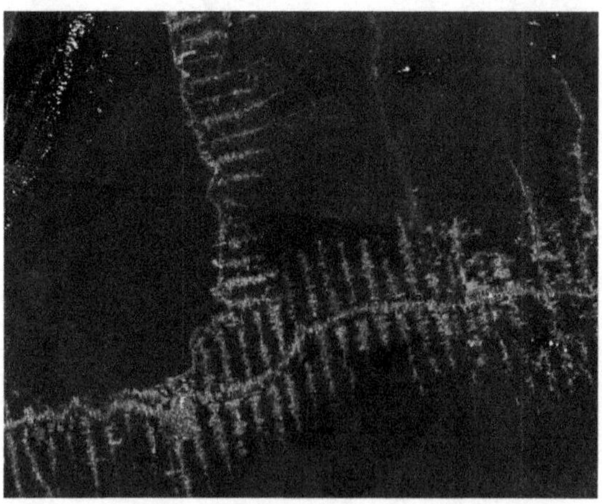

Deforestation in the Amazon rainforest. The “fishbone pattern” is a result of the roads in the forest created by loggers.

Humans place high demand on land not only to build urban centers, but also to build surrounding suburban areas for housing. Land is also allocated for agriculture to sustain the growing population of the city. Expanding cities and suburban areas necessitate corresponding deforestation to meet the land-use and resource requirements of urbanization. Key examples of this are deforestation in the United States and Brazil.

Along with manipulation of land to suit human needs, natural water resources such as rivers and streams are also modified in urban establishments. Modification can come in the form of dams, artificial canals, and even the reversal of rivers. Reversing the flow of the Chicago River is a major example of urban environmental modification.Urban areas in natural desert settings often bring in water from far areas to maintain the human population and will likely have effects on the local desert climate.Modification of aquatic systems in urban areas also results in decreased stream diversity and increased pollution.

Trade, Shipping, and Spread of Invasive Species

A ship navigates through the Firth of Clyde in Scotland, potentially carrying invasive species.

Invasive kudzu vines growing on trees in Atlanta, Georgia, USA

Both local shipping and long-distance trade are required to meet the resource demands important in maintaining urban areas. Carbon dioxide emissions from the transport of goods also contribute to accumulating greenhouse gases and nutrient deposits in the soil and air of urban environments. In addition, shipping facilitates the unintentional spread of living organisms, and introduces them to environments that they would not naturally inhabit. Introduced or alien species are populations of organisms living in a range in which they did not naturally evolve due to intentional or inadvertent human activity. Increased transportation between urban centers furthers the incidental movement of animal and plant species. Alien species often have no natural predators and pose a substantial threat to the dynamics of existing ecological populations in the new environment where they are introduced. Such invasive species are numerous and include house sparrows, ring-necked pheasants, European starlings, brown rats, Asian carp, American bullfrogs, emerald ash borer, kudzu vines, and zebra mussels among numerous others, most notably domesticated animals.

Human Effects on Biogeochemical Pathways

Urbanization results in a large demand for chemical use by industry, construction, agriculture, and energy providing services. Such demands have a substantial impact on biogeochemical cycles, resulting in phenomena such as acid rain, eutrophication, and global warming. Furthermore, natural biogeochemical cycles in the urban environment can be impeded due to impermeable surfaces that prevent nutrients from returning to the soil, water, and atmosphere.

Demand for fertilizers to meet agricultural needs exerted by expanding urban centers can alter chemical composition of soil. Such effects often result in abnormally high concentrations of compounds including sulfur, phosphorus, nitrogen, and heavy metals. In addition, nitrogen and phosphorus used in fertilizers have caused severe problems in the form of agricultural runoff, which alters the concentration of these compounds in local rivers and streams, often resulting in adverse effects on native species. A well-known effect of agricultural runoff is the phenomenon of eutrophication. When the fertilizer chemicals from agricultural runoff reach the ocean, an algal bloom results, then rapidly dies off. The dead algae biomass is decomposed by bacteria that also consume large quantities of oxygen, which they obtain from the water, creating a "dead zone" without oxygen for fish or other organisms. A classic example is the dead zone in the Gulf of Mexico due to agricultural runoff into the Mississippi River.

Just as pollutants and alterations in the biogeochemical cycle alter river and ocean ecosystems, they exert likewise effects in the air. Smog stems from the accumulation of chemicals and pollution and often manifests in urban settings, which has a great impact on local plants and animals. Because urban centers are often considered point sources for pollution, unsurprisingly local plants have adapted to withstand such conditions.

Urban effects on climate

Urban environments and outlying areas have been found to exhibit unique local temperatures, precipitation, and other characteristic activity due to a variety of factors such as pollution and altered geochemical cycles. Some examples of the urban effects on climate are urban heat island, oasis effect, greenhouse gases, and acid rain. This further stirs the debate as to whether urban areas should be considered a unique biome. Despite common trends among all urban centers, the surrounding local environment heavily influences much of the climate. One such example of regional differences can be seen through the urban heat island and oasis effect.

Urban Heat Island Effect

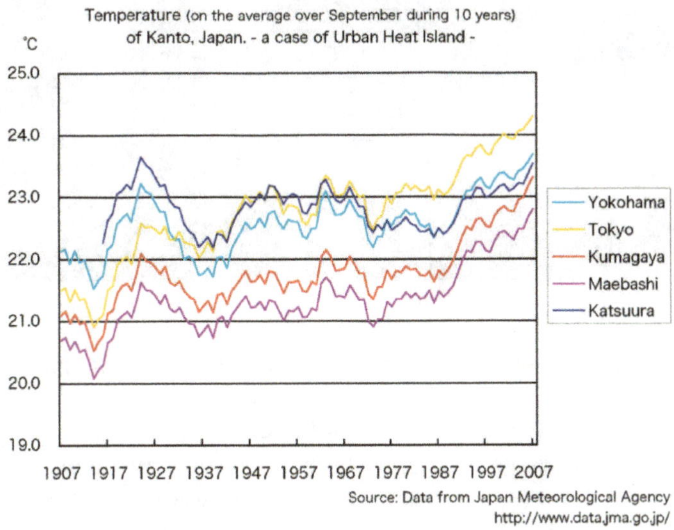

Graphical representation of the rising temperature in Kanto, Japan due to urban heat island.

The urban heat island is a phenomenon in which central regions of urban centers exhibit higher mean temperatures than surrounding urban areas. Much of this effect can be attributed to low city albedo, the reflecting power of a surface, and the increased surface area of buildings to absorb solar radiation. Concrete, cement, and metal surfaces in urban areas tend to absorb heat energy rather than reflect it, contributing to higher urban temperatures. Brazel et al. found that the urban heat island effect demonstrates a positive correlation with population density in the city of Baltimore. The heat island effect has corresponding ecological consequences on resident species. However, this effect has only been seen in temperate climates.

Greenhouse Gases

Greenhouse gas emissions include those of carbon dioxide and methane from the combustion of fossil fuels to supply energy needed by vast urban metropolises. Other greenhouse gases include water vapor, and nitrous oxide. Increases in greenhouse gases due to urban transport, construction, industry and other demands have been correlated strongly with increase in temperature. Sources of methane are agricultural dairy cows and landfills.

Acid rain and polLution

Smokestacks from a wartime production plant releasing pollutants into the atmosphere.

Processes related to urban areas result in the emission of numerous pollutants, which change corresponding nutrient cycles of carbon, sulfur, nitrogen, and other elements. Ecosystems in and around the urban center are especially influenced by these point sources of pollution. High sulfur dioxide concentrations resulting from the industrial demands of urbanization cause rainwater to become more acidic.Such an effect has been found to have a significant influence on locally affected populations, especially in aquatic environments. Wastes from urban centers, especially large urban centers in developed nations, can drive biogeochemical cycles on a global scale.

Urban Environment as an Anthropogenic Biome

The urban environment has been classified as an anthropogenic biome, which is characterized by the predominance of certain species and climate trends such as urban heat island across many urban areas. Examples of species characteristic of many urban environments include, cats, dogs, mosquitoes, rats, flies, and pigeons, which are all generalists. Many of these are dependent on human activity and have adapted accordingly to the niche created by urban centers.

Biodiversity and Urbanization

Research thus far indicates that, on a small scale, urbanization often increases the biodiversity of non-native species while reducing that of native species. This normally results in an overall reduction in species richness and increase in total biomass and species abundance. Urbanization also reduces diversity on a large scale.

Urban stream syndrome is a consistently observed trait of urbanization characterized by high nutrient and contaminant concentration, altered stream morphology, increased dominance of dominant species, and decreased biodiversity The two primary causes of urban stream syndrome are storm water runoff and wastewater treatment plant effluent.

Changes in Diversity

Diversity is normally reduced at intermediate-low levels of urbanization but is always reduced at high levels of urbanization. These effects have been observed in vertebrates and invertebrates while plant species tend to increase with intermediate-low levels of urbanization but these general trends do not apply to all organisms within those groups. For example, McKinney's (2006) review did not include the effects of urbanization on fishes and of the 58 studies on invertebrates, 52 included insects while only 10 included spiders. There is also a geographical bias as most of the studies either took place in North America or Europe.

The effects of urbanization also depend on the type and range of resources used by the organism. Generalist species, those that use a wide range of resources and can thrive under a large range of living conditions, are likely survive in uniform environments. Specialist species, those that use a narrow range of resources and can only cope with a narrow range of living conditions, are unlikely to cope with uniform environments. There will likely be a variable effect on these two groups of organisms as urbanization alters habitat uniformity.

Cause of Diversity Change

The urban environment can decrease diversity through habitat removal and species homogenization - the increasing similarity between two previously distinct biological communities. Habitat degradation and habitat fragmentation reduces the amount of suitable habitat by urban development and separates suitable patches by inhospitable terrain such as roads, neighborhoods, and open parks. Although this replacement of suitable habitat with unsuitable habitat will result in extinctions of native species, some shelter may be artificially created and promote the survival of non-native species (e.g. house sparrow and house mice nests). Urbanization promotes species homogenization through the extinction of native endemic species and the introduction of non-native species that already have a widespread abundance. Changes to the habitat may promote both the extinction of native endemic species and the introduction of non-native species. The effects of habitat change will likely be similar in all urban environments as urban environments are all built to cater to the needs of humans.

The urban environment can also increase diversity in a number of ways. Many foreign organisms are introduced and dispersed naturally or artificially in urban areas. Artificial introductions may be intentional, where organisms have some form of human use, or accidental, where organisms attach themselves to transportation vehicles. Humans provide food sources (e.g. birdfeeder seeds, trash, garden compost) and reduce the numbers of large natural predators in urban environments,allowing large populations to be supported where food and predation would normally limit the population size. There are a variety of different habitats available within the urban environment as a result of differences in land use allowing for more species to be supported than by more uniform habitats.

Civil Engineering and Sustainability

Cities should be planned and constructed in such a way that minimizes the urban effects on the surrounding environment (urban heat island, precipitation, etc.) as well as optimizing ecological activity. For example, increasing the albedo, or reflective power, of surfaces in urban areas, can minimize urban heat island, resulting in a lower magnitude of the urban heat island effect in urban areas. By minimizing these abnormal temperature trends and others, ecological activity would likely be improved in the urban setting.

Need for remediation

Urbanization has indeed had a profound effect on the environment, on both local and global scales. Difficulties in actively constructing habitat corridor and returning biogeochemical cycles to normal raise the question as to whether such goals are feasible. However, some groups are working to return areas of land affected by the urban landscape to a more natural state. This includes using landscape architecture to model natural systems and restore rivers to pre-urban states.

Sustainability

Pipes carrying biogas produced by anaerobic digestion or fermentation of biodegradable materials as a form of carbon sequestration.

With the ever-increasing demands for resources necessitated by urbanization, recent campaigns to move toward sustainable energy and resource consumption, such as LEED certification of buildings, Energy Star certified appliances, and zero emission vehicles, have gained momentum. Sustainability reflects techniques and consumption ensuring reasonably low resource use as a component of urban ecology. Techniques such as carbon recapture may also be used to sequester carbon compounds produced in urban centers rather continually emitting more of the greenhouse gas.

REFERENCES

1. May RM (2001) Stability and Complexity in Model Ecosystems Princeton University Press, reprint of 1973 edition with new foreword. ISBN 978-0-691-08861-7.
2. Pimm SL (2002) Food Webs University of Chicago Press, reprint of 1982 edition with new foreword. ISBN 978-0-226-66832-1.
3. Bolker BM (2008) Ecological models and data in RPrinceton University Press, pages 6–9. ISBN 978-0-691-12522-0.
4. Sugihara G, May R (1990). Nonlinear forecasting as a way of distinguishing chaos from measurement error in time series (PDF). Nature. 344 *(6268): 734–41.* doi:10.1038/344734a0. PMID 2330029.
5. Soetaert K and Herman PMJ (2009) A practical guide to ecological modelling Springer. ISBN 978-1-4020-8623-6.
6. Grant WE (1986) Systems analysis and simulation in wildlife and fisheries sciences. Wiley, University of Minnesota, page 223. ISBN 978-0-471-89236-6.
7. Jopp F (2011) Modeling Complex Ecological DynamicsSpringer, page 122. ISBN 978-3-642-05028-2.
8. Burk AR (2005) New trends in ecology research Nova Publishers, page 136. ISBN 978-1-59454-379-1.
9. Ma T and Wang S (2005) Bifurcation theory and applications World Scientific. ISBN 978-981-256-287-6.
10. May, Robert (1976). Theoretical Ecology: Principles and Applications. Blackwell Scientific Publishers. ISBN 0-632-00768-0.
11. Fisher, R. A. (1930). The genetical theory of natural selection. Oxford: The Clarendon press.
12. R C Lewontin (1961). Evolution and the theory of games. Journal of Theoretical Biology. 1 *(3): 382–403.* doi:10.1016/0022-5193(61)90038-8.
13. John Maynard Smith (1974). Theory of games and evolution of animal conflicts. Journal of Theoretical Biology. 47 *(1): 209–21.* doi:10.1016/0022-5193(74)90110-6. PMID 4459582.
14. Supriatna AK (1998) Optimal harvesting theory for predator–prey metapopulations University of Adelaide, Department of Applied Mathematics.
15. Odum, Eugene P. (1959). Fundamentals of Ecology (Second ed.). Philadelphia and London: W. B. Saunders Co. p. 546 p. ISBN 9780721669410. OCLC 554879.
16. Moss R, Watson A and Ollason J (1982) Animal population dynamics Springer,

page 52–54. ISBN 978-0-412-22240-5.

17. Hal Caswell (2001). Matrix Population Models: Construction, Analysis, and Interpretation. Sinauer.
18. D.T.Crouse, L.B. Crowder, H.Caswell (1987). A stage-based population model for loggerhead sea turtles and implications for conservation. Ecology. 68 *(5): 1412–1423.* doi:10.2307/1939225. JSTOR 1939225.
19. M. Fujiwara; H. Caswell (2001). Demography of the endangered North Atlantic right whale. Nature. 414 *(6863): 537–541.* doi:10.1038/35107054. PMID 11734852.
20. Hubbell, SP (2001). The Unified Neutral Theory of Biodiversity and Biogeography (MPB-32).
21. Bonsall, Michael B.; Hassell, Michael P. (2007). Predator–prey interactions. In May, Robert; McLean, Angela. Theoretical Ecology: Principles and Applications (3rd ed.). Oxford University Press. pp. 46–61.
22. Lotka, A.J., Elements of Physical Biology, Williams and Wilkins, (1925)
23. Goel, N.S. et al., On the Volterra and Other Non-Linear Models of Interacting Populations, Academic Press Inc., (1971)
24. Volterra, V., Variazioni e fluttuazioni del numero d'individui in specie animali conviventi, Mem. Acad. Lincei Roma, 2, 31-113, (1926)
25. Begon, M.; Harper, J. L.; Townsend, C. R. (1988). Ecology: Individuals, Populations and Communities. Blackwell Scientific Publications Inc., Oxford, UK.
26. C.S. Elton (1924). Periodic fluctuations in the numbers of animals - Their causes and effects. Journal of Experimental Biology. 2 *(1): 119–163.*
27. Lipsitch M, Cohen T, Cooper B, Robins JM, Ma S, James L, Gopalakrishna G, Chew SK, Tan CC, Samore MH, Fisman D, Murray M (2003). Transmission dynamics and control of severe acute respiratory syndrome. Science. 300 *(5627): 1966–70. doi:10.1126/science.1086616. PMC 2760158. PMID 12766207.*
28. John D. Reeve; Wiliam W. Murdoch (1986). Biological Control by the Parasitoid Aphytis melinus, and Population Stability of the California Red Scale. Journal of Animal Ecology. 55 *(3): 1069–1082.* doi:10.2307/4434. JSTOR 4434.
29. Arditi, R.; Ginzburg, L.R. (1989). Coupling in predator–prey dynamics: ratio dependence (PDF). Journal of Theoretical Biology. 139 *(3): 311–326.* doi:10.1016/s0022-5193(89)80211-5.
30. Arditi, R. and Ginzburg, L.R. (2012) How Species Interact: Altering the Standard View on Trophic Ecology Oxford University Press. ISBN 9780199913831.
31. Grenfell, Bryan; Keeling, Matthew (2007). Dynamics of infectious disease. In May, Robert; McLean, Angela. Theoretical Ecology: Principles and Applications (3rd ed.). Oxford University Press. pp. 132–147.
32. H. M. Tsuchiya; J. F. Drake; J. L. Jost & A. G. Fredrickson (1972). Predator–Prey Interactions of Dictyostelium discoideum *and Escherichia coli in Continuous Culture1". Journal of Bacteriology.* 110 *(3): 1147–53. PMC 247538. PMID 4555407.*

33. Takeuchi, Y. (1989). Cooperative systems theory and global stability of diffusion models. Acta Applicandae Mathematicae. 14 *(1–2): 49–57.* doi:10.1007/BF00046673.

34. Hubbell, S. P. (2005). The neutral theory of biodiversity and biogeography and Stephen Jay Gould. Paleobiology. 31: 122–123. doi:10.1666/0094-8373(2005)031[0122:TNTOBA]2.0.CO;2.

35. McGill, B. J. (2003). A test of the unified neutral theory of biodiversity. Nature. 422 *(6934): 881–885.* doi:10.1038/nature01583. PMID 12692564.

36. MacArthur RH and Wilson EO (1967) The theory of island biogeography

37. Wiens, J. J.; Donoghue, M. J. (2004). Historical biogeography, ecology and species richness (PDF). Trends in Ecology and Evolution. 19 *(12): 639–644. doi:10.1016/j.tree.2004.09.011. PMID 16701326.*

38. This applies to British and American academics; landscape ecology has a distinct genesis among European academics.

39. First introduced in MacArthur & Wilson›s (1967) book of notable mention in the history and theoretical science of ecology, The Theory of Island Biogeography

40. Reznick, D.; Bryant, M. J.; Bashey, F. (2002). r- and K-selection revisited: The role of population regulation in life-history evolution (PDF). Ecology. 83 *(6): 1509–1520.* doi:10.1890/0012-9658(2002)083[1509:RAKSRT]2.0.CO;2. ISSN 0012-9658.

41. Gewin V (2006). Beyond Neutrality—Ecology Finds Its Niche. PLoS Biol. 4 *(8): 1306–1310. doi:10.1371/journal.pbio.0040278. PMC 1543692. PMID 16895443.*

42. 999) Metapopulation ecology Oxford University Press. ISBN 978-0-19-854065-6.

43. Hanski I, Gilpin M (1991). Metapopulation dynamics: brief history and conceptual domain (PDF). Biological Journal of the Linnean Society. 42: 3–16. doi:10.1111/j.1095-8312.1991.tb00548.x.

44. Cox CB and Moore PD (2010) Biogeography: An Ecological and Evolutionary Approach John Wiley and Sons, page 146. ISBN 978-0-470-63794-4.

45. Vandermeer JH and Goldberg DE (2003) Population ecology: first principles Princeton University Press, page 175–176. ISBN 978-0-691-11441-5.

46. Ilkka Hanski (1982). Dynamics of Regional Distribution: The Core and Satellite Species Hypothesis. Oikos. 38 *(2): 210–221.* doi:10.2307/3544021. JSTOR 3544021.

47. Robert Ulanowicz (). Ecology, the Ascendant Perspective. Columbia Univ. Press. ISBN 0-231-10828-1.

48. Post, D. M. (1993). The long and short of food-chain length. Trends in Ecology and Evolution. 17 *(6): 269–277.* doi:10.1016/S0169-5347(02)02455-2.

49. Jerry Bobrow, Ph.D.; Stephen Fisher (2009). CliffsNotes CSET: Multiple Subjects (2nd ed.). John Wiley and Sons. p. 283. ISBN 978-0-470-45546-3.

50. Elton CS (1927) Animal Ecology. Republished 2001. University of Chicago

Press.

51. Paine RT (1966). Food web complexity and species diversity. The American Naturalist. 100 *(910): 65–75.* doi:10.1086/282400.

52. R.L. Kitching, Systems ecology, University of Queensland Press, 1983, p.9.

53. Bhattacharya K and Vicsek T (2010) Collective decision making in cohesive flocks

54. Li YX, Lukeman R, Edelstein-Keshet L (2007). Minimal mechanisms for school formation in self-propelled particles (PDF). Physica D: Nonlinear Phenomena. 237 *(5): 699–720. doi:10.1016/j.physd.2007.10.009.*

55. Toner J and Tu Y (1995) Long-range order in a two-dimensional xy model: how birds fly together Physical Revue Letters, 75 (23)(1995), 4326–4329.

56. Topaz C, Bertozzi A (2004). Swarming patterns in a two-dimensional kinematic model for biological groups. SIAM J Appl Math. 65 *(1): 152–174.* doi:10.1137/S0036139903437424.

57. Topaz C, Bertozzi A, Lewis M (2006). A nonlocal continuum model for biological aggregation. Bull Math Bio. 68 *(7): 1601–1623.* doi:10.1007/s11538-006-9088-6.

58. Carrillo, J; Fornasier, M; Toscani, G (2010). Particle, kinetic, and hydrodynamic models of swarming (PDF). Modeling and Simulation in Science, Engineering and Technology. 3: 297–336. doi:10.1007/978-0-8176-4946-3_12. ISBN 978-0-8176-4945-6.

59. Smith, Charles H. Wallace›s Unfinished Business. Complexity (publisher Wiley Periodicals, Inc.) Volume 10, No 2, 2004. Retrieved 2007-05-11.

60. Wallace, Alfred. On the Tendency of Varieties to Depart Indefinitely From the Original Type. The Alfred Russel Wallace Page hosted by Western Kentucky University. Retrieved 2007-04-22.

61. Brand, Stewart. For God›s Sake, Margaret. CoEvolutionary Quarterly, June 1976. Retrieved 2007-04-04.

62. Thorp, J. H., & Delong, M. D. (1994). The Riverine Productivity Model: An Heuristic View of Carbon Sources and Organic Processing in Large River Ecosystems. Oikos , 305-308

63. Benke, A. C., Chaubey, I., Ward, G. M., & Dunn, E. L. (2000). Flood Pulse Dynamics of an Unregulated River Floodplain in the Southeastern U.S. Coastal Plain. Ecology , 2730-2741.

64. Cuddington K and Beisner BE (2005) Ecological paradigms lost: routes of theory change Academic Press. ISBN 978-0-12-088459-9.

65. Soulé ME, Simberloff D (1986). What do genetics and ecology tell us about the design of nature reserves? (PDF). Biological Conservation. 35 *(1): 19–40.* doi:10.1016/0006-3207(86)90025-X.

CHAPTER 13

Ecological Effects of Biodiversity

Chapter Contents

Humans are altering the composition of biological communities through a variety of activities that increase rates of species invasions and species extinctions, at all scales, from local to global. These changes in components of the Earth's biodiversity cause concern for ethical and aesthetic reasons, but they also have a strong potential to alter ecosystem properties and the goods and services they provide to humanity. Ecological experiments, observations, and theoretical developments show that ecosystem properties depend greatly on biodiversity in terms of the functional characteristics of organisms present in the ecosystem and the distribution and abundance of those organisms over space and time. Species effects act in concert with the effects of climate, resource availability, and disturbance regimes in influencing ecosystem properties. Human activities can modify all of the above factors; here we focus on modification of these biotic controls.

INTRODUCTION

The diversity of species and genes in ecological communities affects the functioning of these communities. These ecological effects of biodiversity in turn are affected by both climate change through enhanced greenhouse gases, aerosols and loss of land cover, and biological diversity, causing a rapid loss of ecosystems and extinctions of species and local populations. The current rate of extinction is sometimes considered a mass extinction, with current species extinction rates on the order of 100 to 1000 times as high as in the past.

The two main areas where the effect of biodiversity on ecosystem function have been studied are the relationship between diversity and productivity, and the relationship between diversity and community stability. More biologically diverse communities appear to be more productive (in terms of biomass production) than are less diverse communities, and they appear to be more stable in the face of perturbations.

Also animals that inhabit an area may alter the surviving conditions by factors assimilated by climate.

DEFINITIONS

In order to understand the effects that changes in biodiversity will have on ecosystem functioning, it is important to define some terms. Biodiversity is not easily defined, but may be thought of as the number and/or evenness of genes, species, and ecosystems in a region. This definition includes genetic diversity, or the diversity of genes within a species, species diversity, or the diversity of species within a habitat or region, and ecosystem diversity, or the diversity of habitats within a region.

Two things commonly measured in relation to changes in diversity are productivity and stability. Productivity is a measure of ecosystem function. It is generally measured by taking the total aboveground biomass of all plants in an area. Many assume that it can be used as a general indicator of ecosystem function and that total resource use and other indicators of ecosystem function are correlated with productivity.

Stability is much more difficult to define, but can be generally thought of in two ways. General stability of a population is a measure that assumes stability is higher if there is less of a chance of extinction. This kind of stability is generally measured by measuring the variability of aggregate community properties, like total biomass, over time The other definition of stability is a measure of resilience and resistance, where an ecosystem that returns quickly to an equilibrium after a perturbation or resists invasion is thought of as more stable than one that doesn't.

PRODUCTIVITY AND STABILITY AS INDICATORS OF ECOSYSTEM HEALTH

The importance of stability in community ecology is clear. An unstable ecosystem will be more likely to lose species. Thus, if there is indeed a link between diversity and stability, it is likely that losses of diversity could feedback on themselves, causing even more losses of species. Productivity, on the other hand, has a less clear importance in community ecology. In managed areas like cropland, and in areas where animals are grown or caught, increasing productivity increases the economic success of the area and implies that the area has become more efficient, leading to possible long term resource sustainability. It is more difficult to find the importance of productivity in natural ecosystems.

Beyond the value biodiversity has in regulating and stabilizing ecosystem processes, there are direct economic consequences of losing diversity in certain ecosystems and in the world as a whole. Losing species means losing potential foods, medicines, industrial products, and tourism, all of which have a direct economic effect on peoples lives.

Effects on Community Productivity

- **Complementarity** Plant species coexistence is thought to be the result of niche partitioning, or differences in resource requirements among species. By complementarity, a more diverse plant community should be able to use resources more completely, and thus be more productive. Also called niche differentiation, this mechanism is a central principle in the functional group approach, which breaks species diversity down into functional components.
- **Facilitation** : Facilitation is a mechanism whereby certain species help or allow other species to grow by modifying the environment in a way that is favorable to a co-occurring species. Plants can interact through an intermediary like nitrogen, water, temperature, space, or interactions with weeds or herbivores among others. Some examples of facilitation include large desert perennials acting as nurse plants, aiding the establishment of young neighbors of other species by alleviating water and temperature stress, and nutrient enrichment by nitrogen-fixers such as legumes.
- **The Sampling Effect** The sampling effect of diversity can be thought of as having a greater chance of including a species of greatest inherent productivity in a plot that is more diverse. This provides for a composition effect on productivity, rather than diversity being a direct cause. However, the sampling effect may in fact be a compilation of different effects. The sampling effect can be separated into the greater likelihood of selecting a species that is 1) adapted well to particular site conditions, or 2) of a greater inherent productivity. Additionally, one can add to the sampling effect a greater likelihood of including 3) a pair of species that highly complement each other, or 4) a certain species with a large facilitative effect on other members of the community.

Review of Data

Field experiments to test the degree to which diversity affects community productivity have had variable results, but many long term studies in grassland ecosystems have found that diversity does indeed enhance the productivity of ecosystems.Additionally, evidence of this relationship has also been found in grassland microcosms. The differing results between studies may partially be attributable to their reliance on samples with equal species diversities rather than species diversities that mirror those observed in the environment. A 2006 experiment utilizing a realistic variation in species composition for its grassland samples found a positive correlation between increased diversity and increased production.

However, these studies have come to different conclusions as to whether the cause was due more to diversity or to species composition. Specifically, a diversity in the functional roles of the species may be a more important quality for predicting productivity than the diversity in species number. Recent mathematical models have highlighted the importance of ecological context in unraveling this problem. Some models have indicated the importance of disturbance rates and spatial heterogeneity of the environment, others have indicated that the time since disturbance and the habitat's carrying capacity can cause differing relationships. Each ecological context should yield not only a different relationship, but a different contribution to the relationship due to diversity and to composition. The current consensus holds at least that certain combinations of species provide increased community productivity.

Future Research

In order to correctly identify the consequences of diversity on productivity and other ecosystem processes, many things must happen. First, it is imperative that scientists stop looking for a single relationship. It is obvious now from the models, the data, and the theory that there is no one overarching effect of diversity on productivity. Scientists must try to quantify the differences between composition effect and diversity effects, as many experiments never quantify the final realized species diversity (instead only counting numbers of species of seeds planted) and confound a sampling effect for facilitators (a compositional factor) with diversity effects.

Relative amounts of over yielding (or how much more a species grows when grown with other species than it does in monoculture) should be used rather than absolute amounts as relative over yielding can give clues as to the mechanism by which diversity is influencing productivity, however if experimental protocols are incomplete, one may be able to indicate the existence of a complementary or facilitative effect in the experiment, but not be able to recognize its cause. Experimenters should know what the goal of their experiment is, that is, whether it is meant to inform natural or managed ecosystems, as the sampling effect may only be a real effect of diversity in natural ecosystems (managed ecosystems are composed to maximize complementarity and facilitation regardless of species number). By knowing this, they should be able to choose spatial and temporal scales that are appropriate for their experiment. Lastly, to resolve the diversity-function debate, it is advisable that experiments be done with large amounts of spatial and resource heterogeneity and environmental fluctuation over time, as these types of experiments should be able to demonstrate the diversity-function relationship more easily.

Effects on Community Stability

- **Averaging Effect** If all species have differential responses to changes in the ecosystem over time, then the averaging of these responses will cause a more temporally stable ecosystem if more species are in the ecosystem. This effect is a statistical effect due to summing random variables.

- **Negative Covariance Effect** If some species do better when other species are not doing well, then when there are more species in the ecosystem, their overall variance will be lower than if there were fewer species in the system. This lower variance indicates higher stability. This effect is a consequence of competition as highly competitive species will negatively covary.
- **Insurance Effect** If an ecosystem contains more species then it will have a greater likelihood of having redundant stabilizing species, and it will have a greater number of species that respond to perturbations in different ways. This will enhance an ecosystem's ability to buffer perturbations.
- **Resistance to Invasion** Diverse communities may use resources more completely than simple communities because of a diversity effect for complementarity. Thus invaders may have reduced success in diverse ecosystems, or there may be a reduced likelihood that an invading species will introduce a new property or process to a diverse ecosystem.
- **Resistance to Disease** A decreased number of competing plant species may allow the abundances of other species to increase, facilitating the spread of diseases of those species.

Review of Temporal Stability Data

Models have predicted that empirical relationships between temporal variation of community productivity and species diversity are indeed real, and that they almost have to be. Some temporal stability data can be almost completely explained by the averaging effect by constructing null models to test the data against. Competition, which causes negative covariances, only serves to strengthen these relationships.

Review of Resistance and Resilience Stability Data

This area is more contentious than the area of temporal stability, mostly because some have tried generalizing the findings of the temporal stability models and theory to stability in general. While the relationship between temporal variations in productivity and diversity has a mathematical cause, which will allow the relationship to be seen much more often than not, it is not the case with resistance/resilience stability. Some experimenters have seen a correlation between diversity and reduced invasibility, though many have also seen the opposite. The correlation between diversity and disease is also tenuous, though theory and data do seem to support it.

Future Research

In order to more fully understand the effects of diversity on the temporal stability of ecosystems it is necessary to recognize that they are bound to occur. By constructing null models to test the data against (as in Doak et al. 1998) it becomes possible to find situations and ecological contexts where ecosystems become more or less stable than they should be. Finding these contexts would allow for mechanistic studies into why these ecosystems are more stable, which may allow for applications in conservation management.

More importantly more complete experiments into whether diverse ecosystems actually resist invasion and disease better than their less diverse equivalents as invasion and disease are two important factors that lead to species extinctions in the present day.

THEORY AND PRELIMINARY EFFECTS FROM EXAMINING FOOD WEBS

One major problem with both the diversity-productivity and diversity-stability debates discussed up to this point is that both focus on interactions at just a single trophic level. That is, they are concerned with only one level of the food web, namely plants. Other research, unconcerned with the effects of diversity, has demonstrated strong top-down forcing of ecosystems (see keystone species). There is very little actual data available regarding the effects of different food webs, but theory helps us in this area. First, if a food web in an ecosystem has a lot of weak interactions between different species, then it should have more stable populations and the community as a whole should be more stable. If upper levels of the web are more diverse, then there will be less biomass in the lower levels and if lower levels are more diverse they will better be able to resist consumption and be more stable in the face of consumption. Also, top-down forcing should be reduced in less diverse ecosystems because of the bias for species in higher trophic levels to go extinct first. Lastly, it has recently been shown that consumers can dramatically change the biodiversity-productivity-stability relationships that are implied by plants alone. Thus, it will be important in the future to incorporate food web theory into the future study of the effects of biodiversity. In addition this complexity will need to be addressed when designing biodiversity management plans.

Ecological extinction is defined as "the reduction of a species to such low abundance that, although it is still present in the community, it no longer interacts significantly with other species."

Ecological extinction stands out because it is the interaction ecology of a species that is important for conservation work. They state that "unless the species interacts significantly with other species in the community (e.g. it is an important predator, competitor, symbiont, mutualist, or prey) its loss may result in little to no adjustment to the abundance and population structure of other species."

This view stems from the neutral model of communities that assumes there is little to no interaction within species unless otherwise proven.

Estes, Duggins, and Rathburn (1989) recognize two other distinct types of extinction.

Global extinction is defined as "the ubiquitous disappearance of a species."

Local extinction is characterized by "the disappearance of a species from part of its natural range."

KEYSTONE SPECIES

Paine first established the concept of a keystone species by studying the sea star.

Robert Paine (1969) first came up with the concept of a keystone species while studying the effects of the predatory sea star *Pisaster ochraceus*, on the abundance of the herbivorous gastropod, *Tegula funebralis*. This study took place in the rocky intertidal habitat off the coast of Washington; Paine removed all *Pisaster* in 8m x 10m plots weekly while noting the response of *Tegula* for two years. He found that removing the top predator, in this case being *Pisaster*, reduced species number in the treatment plots. Paine defined the concept of a keystone species as a species that has a disproportionate effect on the community structure of an environment in relation to its total biomass. This keystone species effect forms the basis for the concept of ecological extinction.

Examples

Sea otters maintain the overall biodiversity of the kelp forest community.

Estes et al. (1978) evaluated the potential role of the sea otter as the keystone predator in near-shore kelp forests. They compared the Rat and Near islands in the Aleutian islands to test if "sea otter predation controls epibenthic invertebrate populations (specifically sea urchins), and in turn releases the vegetation association from intense grazing". Estes and his colleagues found that different size structures and densities of sea urchins were correlated with the presence of sea otter populations, and because they are the principal prey of this keystone predator, the sea otters were most likely the main determinants of the differences in sea urchin populations. With high sea otter densities the herbivory of sea urchins in these kelp forest was severely limited, and this made competition between algal species the main determinant in survival. However, when sea otters were absent, herbivory of the sea urchins was greatly intensified to the point of decimation of the kelp forest community. This loss of heterogeneity serves as a loss of habitat for both fish and eagle populations that depend on the richly productive kelp forest environment. Historical over harvesting of sea otter furs has severely restricted their once wide-ranging habitat, and only today are scientists starting to see the implications of these local extinctions. Conservation work needs to focus on finding the density threshold that render the sea otters an effective population. It must then continue and artificially repopulate the historical range of the sea otter in order to allow kelp forest communities to re-establish.

The California spiny lobster, or *Panulirus interruptus*, is another example of a keystone predator that has a distinct role in maintaining species diversity in its habitat. Robles (1987) demonstrated experimentally that the exclusion of spiny lobsters from the intertidal zone habitats led to the competitive dominance of mussels (*Mytilus edulis* and *M. californianus*). This results shows another example of how the ecological extinction of a keystone predator can reduce species diversity in an ecosystem. Unfortunately, the threshold of ecological extinction has long passed due to over fishing now that many local extinctions of the California spiny lobster are common.

Jackson et al. (2001) took a much needed historical perspective on the role of ecological extinction caused by overfishing of oysters in the Chesapeake Bay. Commercial oyster fishing had not affected the bay ecosystem until mechanical dredges for harvesting were utilized in the 1870s. The bay today is plagued by eutrophication due to algal blooms, and the resulting water is highly hypoxic. These algal blooms have competitively excluded any other species from surviving, including the rich diversity in faunal life that once flourished such as dolphins, manatees, river otters, sea turtles, alligators, sharks, and rays. This example highlights the top-down loss of diversity commercial fishing has on marine ecosystems by removing the keystone species of the environment.

Invasive Species

Novaro et al. (2000) assessed the potential ecological extinction of guanacos (*Lama guanocoe*) and lesser rheas (*Pterocnemia pennata*) as a prey source for native omnivores and predators in the Argentine Patagonia. These native species are being replaced by introduced species such as the European rabbit, red deer, and domestic cattle; the cumulative damage from the increased herbivory by introduced species has also served to accelerate destruction of the already dwindling Argentine pampas and steppe habitats. This was the first study to take into account a large number of diverse predators, ranging from skunks to pumas, as well as conduct their survey in non-protected areas that represent the majority of southern South America. Novaro and his colleagues found that the entire assemblage of native carnivores relied primarily on introduced species as a prey base. They also suggested that the lesser rhea and guanaco had already passed their ecological effective density as a prey species, and thus were ecological extinct. It is possible that the niches of introduced species as herbivores too closely mirrored those of the natives, and thus competition was the primary cause of ecological extinction. The effect of introduction of new competitors, such as the red deer and rabbit, also served to alter the vegetation in the habitat, which could have further pronounced the intensity of competition. Guanacos and rheas have been classified as a low risk for global extinction, but this simplistic view of their demography doesn't take into account that they have already become functionally extinct in the Argentine Patagonia. Novaro and his colleagues suggest «this loss could have strong effects on plant-animal interactions, nutrient dynamics, and disturbance regimes ...» This is a prime example of how current conservation policy has already failed to protect the intended species because of its lack of a functionally sound definition for extinction.

Christian found Argentine ants to disrupt large seed dispersal mutualisms.

Seed dispersal mechanisms play a fundamental role in the regeneration and continuation of community structure, and a recent study by Christian (2001) demonstrated a shift in the composition of the plant community in the South African shrublands following an invasion by the Argentine ant *(Linepithema humile)*. Ants disperse up to 30% of the flora in the shrublands and are vital to the survival of fynbos plants because they bury the large seeds away from the dangers of predation and fire damage. It is also crucial for seeds to be buried, because nearly all seed germination takes place in the first season after a fire. Argentine ants, a recent invader, do not disperse even small seeds. Christian tested whether the invasion of the Argentine ant differentially effected small and large-seeded fauna. He found that post-fire recruitment of large-seeded flora was reduced disproportionately for large seeds in sites already invaded by Argentine ants. These initial low large-seed density recruitments will eventually lead the domination of small-seeded fauna in invaded habitats. The consequences of this change in community structure highlight the struggle for dispersal of large-seeded flora that have potential reverberations around the world because ants are major ecological seed dispersers throughout the globe.

Modeling Ecological Extinction

Below a certain density threshold, flying foxes are no longer effective seed dispersers.

The McConkey and Drake (2006) study is unique because it was one of the first attempts to model a density-dependent threshold relationship that described ecological extinction. They studied a seed dispersal interaction between flying foxes and trees with large seeds on the tropical Pacific Islands. Insular flying foxes (*Pteropus tonganus)*, are considered to be keystone species because they are the only seed dispersers that can carry large seeds long distances. The host-pathogen model by Janzen and Connell suggests that survivorship of seeds in the tropics greatly increases the further away from the parent tree it lands, and that trees require this dispersal in order to avoid extinction. In the pathogen latent environment of the tropics, seed dispersal only becomes more paramount to species survival. As hypothesized, McConkey and Drake found a threshold relationship between the Flying Fox Index (FFI) and the median proportion of seeds carried over five meters. Below the threshold of abundance seed dispersal was insignificant and independent of flying fox abundance; however, above the threshold, dispersal positively correlated with increased flying fox abundance (as measured by the FFI). Although they did not directly prove the cause for this relationship, McConkey and Drake proposed a behavioral mechanism. Flying foxes are known to be territorial, and in the absence of competition a flying fox will eat within one tree, effectively dropping the seeds right below it. Alternatively, if there is a high density of flying foxes feeding at one time (abundance above the threshold density) then aggressive behavior, such as stealing fruit from another individual's territory, will lead to longer average seed dispersal. In this way the seed dispersing flying fox has a disproportional effect on the overall community structure in comparison to their relative biomass. Modeling the effect of ecological extinction on communities is the first step to applying this framework into conservation work.

While ecologists are just starting to get a grapple on the significant interactions within an ecosystem, they must continue to find an effective density threshold that can maintain the level of equilibrium species diversity. Only with this knowledge of where and to what extent a specific species interacts with its environment will the proper and most efficient levels of conservation work take place. This work is especially important on the limited ecosystems of islands, where there are less likely to be replacement species for specific niches. With species diversity and available habitat decreasing rapidly worldwide, identifying the systems that are most crucial to the ecosystem will be the crux of conservation work.

Climate Change

Climate change has produced numerous shifts in the distributions and abundances of species. Thomas et al. (2004) went on to assess the extinction risk due to these shifts over a broad range of global habitats. Their predictive model using midline estimates for climate warming over the next 50 years suggests that 15-37% of species will be "committed to extinction" be 2050. Although the average global temperature has risen .6°C, individual populations and habitats will only respond to their local changes in climate. Root et al. (2002) suggests that local changes in climate may account for density changes in regions, shifts in phenology (timing) of events, changes in morphology (biology) (such as body size), and shifts in genetic frequencies. They found that there have been an average phenological shift of 5.1 days earlier in the spring for a broad range of over a thousand compiled studies. This shift was also, as predicted, more pronounced in the upper latitudes that have concurrently had the largest shift in local average temperatures.

While the loss of habitat, loss of pollinator mutualisms, and the effect of introduced species all have distinct pressures on native populations, these effects must be looked underneath a synergistic and not an independent framework. Climate change has the potential to exacerbate all of these processes. Nehring (1999) found a total of 16 non-indigenous thermophilic phytoplankton established in habitats northwards of their normal range in the North Sea. He likened these changes in range of more southerly phytoplankton to climatic shifts in ocean temperature. All of these effects have additive effects to the stress on populations within an environment, and with the additionally fragile and more complete definition of ecological extinction must be taken into account into preventative conservation measures.

Implications for Conservation Policy

Conservation policy has historically lagged behind current science all over the world, but at this critical juncture politicians must make the effort to catch up before massive extinctions occur on our planet. For example, the pinnacle of American conservation policy, the Endangered Species Act of 1973, fails to acknowledge any benefit for protecting highly interactive species that may help maintain overall species diversity. Policy must first assess whether the species in question is considered highly interactive by asking the questions "does the absence or loss of this species, either directly or indirectly, incur a loss of overall diversity, effect the reproduction or recruitment of other species, lead to a change in habitat structure, lead to a change in productivity or nutrient dynamics between ecosystems, change important ecological processes, or reduce the resilience of the ecosystem to disturbances?". After these multitudes of questions are addressed to define an interactive species, an ecologically effective density threshold must be estimated in order to maintain this interaction ecology. This process holds many of the same variables contained within viable population estimates, and thus should not be difficult to incorporate into policy. To avoid mass extinction on a global scale unlike anyone has seen before, scientists must understand all of the mechanisms driving the process. It is now that the governments of the world must act in order to prevent this catastrophe of the loss of biodiversity from progressing further and wasting all of the time and money spent on previous conservation efforts.

ENDEMISM

The orange-breasted sunbird (*Nectarinia violacea*) is exclusively found in fynbos vegetation.

Bicolored frog (*Clinotarsus curtipes*) is endemic to the Western Ghats of India

Endemism is the ecological state of a species being unique to a defined geographic location, such as an island, nation, country or other defined zone, or habitat type; organisms that are indigenous to a place are not endemic to it if they are also found elsewhere. The extreme opposite of endemism is cosmopolitan distribution. An alternative term for a species that is endemic is **precinctive**, which applies to species (and subspecific categories) that are restricted to a defined geographical area.

The word *endemic* is from New Latin *endēmicus*, from Greek ενδήμος, *endēmos*, "native." *Endēmos* is formed of *en* meaning "in," and *dēmos* meaning "the people." The term, *precinctive*, has been suggested by some scientists, and was first used in botany by MacCaughey in 1917. It is the equivalent of «endemism». *Precinction* was perhaps first used by Frank and McCoy. *Precinctive* seems to have been coined by David Sharp of the Hawaiian fauna in 1900: "I use the word precinctive in the sense of ‹confined to the area under discussion› ... ‹precinctive forms› means those forms that are confined to the area specified.» That definition excludes artificial confinement of examples by humans in far-off botanical gardens or zoological parks.

Physical, climatic, and biological factors can contribute to endemism. The orange-breasted sunbird is exclusively found in the fynbos vegetation zone of southwestern South Africa. The glacier bear is found only in limited places in Southeast Alaska. Political factors can play a part if a species is protected, or actively hunted, in one jurisdiction but not another.

There are two subcategories of endemism: paleoendemism and neoendemism. Paleoendemism refers to species that were formerly widespread but are now restricted to a smaller area. Neoendemism refers to species that have recently arisen, such as through divergence and reproductive isolation or through hybridization and polyploidy in plants.

Endemic types or species are especially likely to develop on geographically and biologically isolated areas such as islands and remote island groups, such as Hawaii, the Galápagos Islands, and Socotra; they can equally develop in biologically isolated areas such as the highlands of Ethiopia, or large bodies of water far from other lakes, like Lake Baikal.

Endemics can easily become endangered or extinct if their restricted habitat changes, particularly—but not only—due to human actions, including the introduction of new organisms. There were millions of both Bermuda petrels and "Bermuda cedars» (actually *junipers*) in Bermuda when it was settled at the start of the seventeenth century. By the end of the century, the petrels were thought extinct. Cedars, already ravaged by centuries of shipbuilding, were driven nearly to extinction in the twentieth century by the introduction of a parasite. Bermuda petrels and cedars are now rare, as are other species endemic to Bermuda.

Threats to Highly Endemistic Regions

Principal causes of habitat degradation and loss in highly endemistic ecosystems include agriculture, urban growth, surface mining, mineral extraction, logging operations and slash-and-burn agriculture.

INTRODUCED SPECIES

Sweet clover (*Melilotus sp.*), introduced and naturalized to the U.S. from Eurasia as a forage and cover crop.

An introduced, alien, exotic, non-indigenous, or non-native species, or simply an introduction, is a species living outside its native distributional range, which has arrived there by human activity, either deliberate or accidental. Non-native species can have various effects on the local ecosystem. Introduced species that become established and spread beyond the place of introduction are called invasive species. Some have a negative effect on a local ecosystem. Some introduced species may have no negative effect or only minor impact. Some species have been introduced intentionally to combat pests. They are called biocontrols and may be regarded as beneficial as an alternative to pesticides in agriculture for example. In some instances the potential for being beneficial or detrimental in the long run remains unknown. A list of some introduced species is given in a separate article.

The effects of introduced species on natural environments have gained much scrutiny from scientists, governments, farmers and others.

Terminology

The terminology associated with introduced species is now in flux for various reasons. Other terms with somewhat similar meanings) with *introduced* are *acclimatized*, *adventive*, *naturalized*, and *immigrant* but those terms refer to a subset of introduced species: those that have become established and can reproduce without human assistance. The term invasive refers only to those species that become established and spread beyond the place of introduction. For practical purposes, this term is applied only to invasive species that cause damage.

In the broadest and most widely used sense, an introduced species is synonymous with *non-native* and therefore applies as well to most garden and farm organisms; these adequately fit the basic definition given above. However, some sources add to that basic definition "and are now reproducing in the wild," which removes from consideration as *introduced* all of those species raised or grown in gardens or farms that do not survive without tending by people. With respect to plants, these latter are in this case defined as either *ornamental* or *cultivated* plants.

The following definition from the United States Environmental Protection Agency (EPA), although perhaps lacking ecological sophistication, is more typical: *introduced species* "A species that has been intentionally or inadvertently brought into a region or area. Also called an exotic or non-native species." Introduction of a species outside its native range is all that is required to be qualified as an "introduced species" such that one can distinguish between introduced species that may not occur except in cultivation, under domestication or captivity whereas others become established outside their native range and reproduce without human assistance. Such species might be termed "naturalized", "established", "wild non-native species". If they further spread beyond the place of introduction they are called "invasive". The transition from introduction, to establishment and to invasion has been described in the context of plants. Introduced species are essentially «non-native» species. Invasive species are those introduced species that spreadwidely or quickly and cause harm, be that to the environment, human health, other valued resources or the economy. There have been calls from scientists to consider a species «invasive» only in terms of their spread and reproduction rather than the harm they may cause.

According to a practical definition, an invasive species is one that has been introduced and become a pest in its new location, spreading (invading) by natural means. The term is used to imply both a sense of urgency and actual or potential harm. For example, U.S. Executive Order 13112 (1999) defines "invasive species" as "an alien species whose introduction does or is likely to cause economic or environmental harm or harm to human health". The biological definition of invasive species, on the other hand, makes no reference to the harm they may cause, only to the fact that they spread beyond the area of original introduction.

Although some argue that «invasive» is a loaded word and harm is difficult to define, the fact of the matter is that organisms have and continue to be introduced to areas in which they are not native, sometimes with but usually without much regard to the harm that could result.

From a regulatory perspective, it is neither desirable nor practical to list as undesirable or outright ban all non-native species (although the State of Hawaii has adopted an approach that comes close to this). Regulations require a definitional distinction between non-natives that are deemed especially onerous and all others. Introduced *pest species* that are *officially listed* as invasive, best fit the definition of an *invasive species*. Early detection and rapid response is the most effective strategy for regulating a pest species and reducing economic and environmental impacts of an introduction

In Great Britain, the Wildlife and Countryside Act 1981 prevents the introduction of any animal not naturally occurring in the wild or any of a list of both animals or plants introduced previously and proved to be invasive.

Nature of Introductions

By definition, a species is considered "introduced" when its transport into an area outside of its native range is human mediated. Introductions by humans can be described as either intentional or accidental. Intentional introductions have been motivated by individuals or groups who either (1) believe that the newly introduced species will be in some way beneficial to humans in its new location or, (2) as is the case with pythons in the Everglades, species are introduced intentionally but with no regard to the potential impact. Unintentional or accidental introductions are most often a byproduct of human movements, and are thus unbound to human motivations. Subsequent range expansion of introduced species may or may not involve human activity.

Intentional Introductions

Species that humans intentionally transport to new regions can subsequently become successfully established in two ways. In the first case, organisms are purposely released for establishment in the wild. It is sometimes difficult to predict whether a species will become established upon release, and if not initially successful, humans have made repeated introductions to improve the probability that the species will survive and eventually reproduce in the wild. In these cases it is clear that the introduction is directly facilitated by human desires.

Male *Lophura nycthemera* (silver pheasant), a native of East Asia that has been introduced into parts of Europe for ornamental reasons.

In the second case, species intentionally transported into a new region may escape from captive or cultivated populations and subsequently establish independent breeding populations. Escaped organisms are included in this category because their initial transport to a new region is human motivated.

Perhaps the most common motivation for introducing a species into a new place is that of economic gain. Examples of species introduced for the purposes of benefiting agriculture, aquaculture or other economic activities are widespread.Eurasian carp was first introduced to the United States as a potential food source. The apple snail was released in Southeast Asia with the intent that it be used as a protein source, and subsequently to places like Hawaii to establish a food industry. In Alaska, foxes were introduced to many islands to create new populations for the fur trade. About twenty species of African and European dung beetles have established themselves in Australia after deliberate introduction by the Australian Dung Beetle Project in an effort to reduce the impact of livestock manure. The timber industry promoted the introduction of Monterey pine (*Pinus radiata*) from California to Australia and New Zealand as a commercial timber crop. These examples represent only a small subsample of species that have been moved by humans for economic interests.

Introductions have also been important in supporting recreation activities or otherwise increasing human enjoyment. Numerous fish and game animals have been introduced for the purposes of sport fishing and hunting (Earthworms as invasive species). The introduced amphibian (*Ambystoma tigrinum*) that threatens the endemic California salamander (*Ambystoma californiense*) was introduced to California as a source of bait for fishermen. Pet animals have also been frequently transported into new areas by humans, and their escapes have resulted in several successful introductions, such as those of feral cats and parrots.

Many plants have been introduced with the intent of aesthetically improving public recreation areas or private properties. The introduced Norway maple for example occupies a prominent status in many of Canada's parks. The transport of ornamental plants for landscaping use has and continues to be a source of many introductions. Some of these species have escaped horticultural control and become invasive. Notable examples include water hyacinth, salt cedar, and purple loosestrife.

In other cases, species have been translocated for reasons of «cultural nostalgia,» which refers to instances in which humans who have migrated to new regions have intentionally brought with them familiar organisms. Famous examples include the introduction of starlings to North America by Englishman Eugene Schieffelin, a lover of the works of Shakespeare and the chairman of the American Acclimatization Society, who, it is rumoured, wanted to introduce all of the birds mentioned in Shakespeare›s plays into the United States. He deliberately released eighty starlings into Central Park in New York City in 1890, and another forty in 1891.

Yet another prominent example is the introduction of the European rabbit to Australia by one Thomas Austin, a British landowner who had the rabbits released on his estate in Victoria because he missed hunting them. A more recent example is the introduction of the common wall lizard to North America by a Cincinnati boy, George Rau, around 1950 after a family vacation to Italy.

Intentional introductions have also been undertaken with the aim of ameliorating environmental problems. A number of fast spreading plants such as garlic mustard and kudzu have been introduced as a means of erosion control. Other species have been introduced as biological control agents to control invasive species and involves the purposeful introduction of a natural enemy of the target species with the intention of reducing its numbers or controlling its spread.

A special case of introduction is the reintroduction of a species that has become locally endangered or extinct, done in the interests of conservation. Examples of successful reintroductions include wolves to Yellowstone National Park in the U.S., and the red kite to parts of England and Scotland. Introductions or translocations of species have also been proposed in the interest of genetic conservation, which advocates the introduction of new individuals into genetically depauperate populations of endangered or threatened species.

The above examples highlight the intent of humans to introduce species as a means of incurring some benefit. While these benefits have in some cases been realized, introductions have also resulted in unforeseen costs, particularly when introduced species take on characteristics of invasive species.

Non-native species can become such a common part of an environment, culture, and even diet that little thought is given to their geographic origin. For example, soybeans, kiwi fruit, wheat and all livestock except the American bison and the turkey are non-native species to North America. Collectively, non-native crops and livestock comprise 98% of US food. These and other benefits from non-natives are so vast that, according to the Congressional Research Service, they probably exceed the costs.

Unintentional Introductions

Unintentional introductions occur when species are transported by human vectors. For example, three species of rat (the black, Norway and Polynesian) have spread to most of the world as hitchhikers on ships. There are also numerous examples of marine organisms being transported in ballast water, one being the zebra mussel. Over 200 species have been introduced to the San Francisco Bay in this manner making it the most heavily invaded estuary in the world. Increasing rates of human travel are providing accelerating opportunities for species to be accidentally transported into areas in which they are not considered native. There is also the accidental release of the Africanized honey bees (AHB), known colloquially as «killer bees» or Africanized bee to Brazil in 1957 and the Asian carps to the United States. The insect commonly known as the brown marmorated stink bug (*Halyomorpha halys*) was introduced accidentally in Pennsylvania. Another form of unintentional introductions is when an intentionally introduced plant carries a parasite or herbivore with it. Some become invasive, for example the oleander aphid, accidentally introduced with the ornamental plant, oleander.

Introduced Plants

Many non-native plants have been introduced into new territories, initially as either ornamental plants or for erosion control, stock feed, or forestry. Whether an exotic will become an invasive species is seldom understood in the beginning, and many non-native ornamentals languish in the trade for years before suddenly naturalizing and becoming invasive.

Peaches, for example, originated in China, and have been carried to much of the populated world. Tomatoes are native to the Andes. Squash (pumpkins), maize (corn), and tobacco are native to the Americas, but were introduced to the Old World. Many introduced species require continued human intervention to survive in the new environment. Others may become feral, but do not seriously compete with natives, but simply increase the biodiversity of the area.

Dandelions are also introduced species to North America.

A very troublesome marine species in southern Europe is the seaweed *Caulerpa taxifolia*. *Caulerpa* was first observed in the Mediterranean Sea in 1984, off the coast of Monaco. By 1997, it had covered some 50 km^2. It has a strong potential to overgrow natural biotopes, and represents a major risk for sublittoral ecosystems. The origin of the alga in the Mediterranean was thought to be either as a migration through the Suez Canal from the Red Sea, or as an accidental introduction from an aquarium. Another troublesome plant species is the terrestrial plant *Phyla canescens,* which was intentionally introduced into many countries in North America, Europe, and Africa as an ornamental plant. This species has become invasive in Australia, where it threatens native rare plants and causes erosion and soil slumping around river banks. It has also become invasive in France where it has been listed as an invasive plant species of concern in the Mediterranean region, where it can form monocultures that threaten critical conservation habitats.

Japanese knotweed grows profusely in many nations. Human beings introduced it into many places in the 19th century. It is a source of resveratrol, a dietary supplement.

Introduced Animals

Bear in mind that most introduced species do not become invasive. Examples of introduced animals that have become invasive include the gypsy moth in eastern North America, the zebra mussel and alewife in the Great Lakes, the Canada goose and gray squirrel in Europe, the muskrat in Europe and Asia, the cane toad and red fox in Australia, nutria in North America, Eurasia, and Africa, and the common brushtail possum in New Zealand.

Most Commonly Introduced Species

Some species, such as the brown rat, house sparrow, ring-necked pheasant and European starling, have been introduced very widely. In addition there are some agricultural and pet species that frequently become feral; these include rabbits, dogs, goats, fish, pigs and cats.

Invasive exotic diseases

History is rife with the spread of exotic diseases, such as the introduction of smallpox into the indigenous peoples of the Americas by the Spanish, where it obliterated entire populations of indigenous civilizations before they were ever even seen by Europeans.

Problematic exotic disease introductions in the past century or so include the chestnut blight which has almost eliminated the American chestnut tree from its forest habitat, and Dutch elm disease, which has severely reduced the American elm trees in forests and cities.

Diseases may also be vectored by invasive insects such as the Asian citrus psyllid and the bacterial disease citrus greening.

Introduced Species on Islands

Perhaps the best place to study problems associated with introduced species is on islands. Depending upon the isolation (how far an island is located from continental biotas), native island biological communities may be poorly adapted to the threat posed by exotic introductions. Often this can mean that no natural predator of an introduced species is present, and the non-native spreads uncontrollably into open or occupied niche.

An additional problem is that birds native to small islands may have become flightless because of the absence of predators prior to introductions and cannot readily escape danger. The tendency of rails in particular to evolve flightless forms on islands has led to the disproportionate number of extinctions in that family.

The field of island restoration has developed as a field of conservation biology and ecological restoration, a great deal of which deals with the eradication of introduced species.

New Zealand

In New Zealand the largest commercial crop is *Pinus radiata*, the native Californian Monterey pine tree, which grows as well in New Zealand as in California. However, the pine forests are also occupied by deer from North America and Europe and by possums from Australia. All are exotic species and all have thrived in the New Zealand environment. The pines are seen as beneficial while the deer and possums are regarded as serious pests.

Common gorse, originally a hedge plant in Britain, was introduced to New Zealand for the same purpose. Like the Monterey pine, it has shown a favour to its new climate. It is, however, regarded as a noxious plant that threatens to obliterate native plants in much of the country and is hence routinely eradicated, though it can also provide a nursery environment for native plants to reestablish themselves.

Rabbits, introduced as a food source by sailors in the 1800s, have become a severe nuisance to farmers, notably in the South Island. The myxomatosis virus was illegally imported and illegally released, but it had little lasting effect upon the rabbit population other than to make it more resistant to the virus.

Cats, brought by the Māori and later by Europeans, have had a devastating effect upon the native birdlife, particularly as many New Zealand birds are flightless. Feral cats and dogs which were originally brought as pets are also known to kill large numbers of birds. A recent (2006) study in the South Island has shown that even domestic cats with a ready supply of food from their owners may kill hundreds of birds in a year, including natives.

Sparrows, which were brought to control insects upon the introduced grain crops, have displaced native birds as have rainbow lorikeets and cockatoos (both from Australia) which fly free around areas west of Auckland City such as the Waitakere Ranges.

In much of New Zealand, the Australian black swan has effectively eliminated the existence of the previously introduced mute swan.

Two notable varieties of spiders have also been introduced: the white tail spider and the redback spider. Both may have arrived inside shipments of fruit. Until then, the only spider (and the only poisonous animal) dangerous to humans was the native katipo, which is very similar to the redback and interbreed with the more aggressive Australian variety.

Introduced Species on a Planetary Body

It has been hypothesized that invasive species of microbial life could contaminate a planetary body after the former is introduced by a space probe or spacecraft, either deliberately or unintentionally.

LATITUDINAL GRADIENTS IN SPECIES DIVERSITY

The increase in species richness or biodiversity that occurs from the poles to the tropics, often referred to as the latitudinal diversity gradient (LDG), is one of the most widely recognized patterns in ecology. Put another way, in the present day localities at lower latitudes generally have more species than localities at higher latitudes. The LDG has been observed to varying degrees in Earth›s past.

Explaining the latitudinal diversity gradient is one of the great contemporary challenges of biogeography and macroecology (Willig et al. 2003, Pimm and Brown 2004, Cardillo et al. 2005). The question "What determines patterns of species diversity?" was among the 25 key research themes for the future identified in 125th Anniversary issue of *Science* (July 2005). There is a lack of consensus among ecologists about the mechanisms underlying the pattern, and many hypotheses have been proposed and debated. A recent review noted that among the many conundrums associated with the LDG (or LBG, Latitudinal Biodiversity Gradient) the causal relationship between rates of molecular evolution and speciation has yet to be demonstrated.

Understanding the global distribution of biodiversity is one of the most significant objectives for ecologists and biogeographers. Beyond purely scientific goals and satisfying curiosity, this understanding is essential for applied issues of major concern to humankind, such as the spread of invasive species, the control of diseases and their vectors, and the likely effects of global climate change on the maintenance of biodiversity (Gaston 2000). Tropical areas play a prominent role in the understanding of the distribution of biodiversity, as their rates of habitat degradation and biodiversity loss are exceptionally high.

Patterns in the Past

The LDG is a noticeable pattern among modern organisms that has been described qualitatively and quantitatively. It has been studied at various taxonomic levels, through different time periods and across many geographic regions (Crame 2001). The LDG has been observed to varying degrees in Earth's past, possibly due to differences in climate during various phases of Earth's history. Some studies indicate that the LDG was strong, particularity among marine taxa, while other studies of terrestrial taxa indicate the LDG had little effect on the distribution of animals.

Hypotheses for Pattern

Although many of the hypotheses exploring the latitudinal diversity gradient are closely related and interdependent, most of the major hypotheses can be split into three general hypotheses.

Spatial/Area Hypotheses

There are five major hypotheses that depend solely on the spatial and areal characteristics of the tropics.

Mid-domain Effect

Using computer simulations, Colwell and Hurtt (1994) and Willig and Lyons (1998) first pointed out that if species' latitudinal ranges were randomly shuffled within the geometric constraints of a bounded biogeographical *domain* (e.g. the continents of the New World, for terrestrial species), species' ranges would tend to overlap more toward the center of the domain than towards its limits, forcing a mid-domain peak in species richness. Colwell and Lees (2000) called this stochastic phenomenon the *mid-domain effect* (MDE), presented several alternative analytical formulations for one-dimensional MDE (expanded by Connolly 2005), and suggested the hypothesis that MDE might contribute to the latitudinal gradient in species richness, together with other explanatory factors considered here, including climatic and historical ones. Because «pure» mid-domain models attempt to exclude any direct environmental or evolutionary influences on species richness, they have been claimed to be null models (Colwell et al. 2004, 2005). On this view, if latitudinal gradients of species richness were determined solely by MDE, observed richness patterns at the biogeographic level would not be distinguishable from patterns produced by random placement of observed ranges (Colwell and Lees 2000). Others object that MDE models so far fail to exclude the role of environment at the population level and in setting domain boundaries, and therefore cannot be considered null models (Hawkins and Diniz-Filho 2002; Hawkins et al. 2005; Zapata et al. 2003, 2005). Mid-domain effects have proven controversial (e.g. Jetz and Rahbek 2001, Koleff and Gaston 2001, Lees and Colwell, 2007, Romdal et al. 2005, Rahbek et al. 2007, Storch et al. 2006; Bokma and Monkkonen 2001, Diniz-Filho et al. 2002, Hawkins and Diniz-Filho 2002, Kerr et al. 2006, Currie and Kerr, 2007). While some studies have found evidence of a potential role for MDE in latitudinal gradients of species richness, particularly for wide-ranging species (e.g. Jetz and Rahbek 2001, Koleff and Gaston 2001, Lees and Colwell, 2007, Romdal et al. 2005, Rahbek et al. 2007, Storch et al. 2006; Dunn et al. 2007) others report little correspondence between predicted and observed latitudinal diversity patterns (Bokma and Monkkonen 2001, Currie and Kerr, 2007, Diniz-Filho et al. 2002, Hawkins and Diniz-Filho 2002, Kerr et al. 2006).

Geographical area Hypothesis

Another spatial hypothesis is the geographical area hypothesis (Terborgh 1973). It asserts that the tropics are the largest biome and that large tropical areas can support more species.

More area in the tropics allows species to have larger ranges, and consequently larger population sizes. Thus, species with larger ranges are likely to have lower extinction rates (Rosenzweig 2003). Additionally, species with larger ranges may be more likely to undergo allopatric speciation, which would increase rates of speciation (Rosenzweig 2003). The combination of lower extinction rates and high rates of speciation leads to the high levels of species richness in the tropics.

A critique of the geographical area hypothesis is that even if the tropics is the most extensive of the biomes, successive biomes north of the tropics all have about the same area. Thus, if the geographical area hypothesis is correct these regions should all have approximately the same species richness, which is not true, as is referenced by the fact that polar regions contain fewer species than temperate regions (Gaston and Blackburn 2000). To explain this, Rosenzweig (1992) suggested that if species with partly tropical distributions were excluded, the richness gradient north of the tropics should disappear. Blackburn and Gaston 1997 tested the effect of removing tropical species on latitudinal patterns in avian species richness in the New World and found there is indeed a relationship between the land area and the species richness of a biome once predominantly tropical species are excluded. Perhaps a more serious flaw in this hypothesis is some biogeographers suggest that the terrestrial tropics are not, in fact, the largest biome, and thus this hypothesis is not a valid explanation for the latitudinal species diversity gradient (Rohde 1997, Hawkins and Porter 2001). In any event, it would be difficult to defend the tropics as a "biome" rather than the geographically diverse and disjunct regions that they truly include.

The effect of area on biodiversity patterns has been shown to be scale dependent, having the strongest effect among species with small geographical ranges compared to those species with large ranges who are affected more so by other factors such as the mid-domain and/or temperature.

Species-Energy Hypothesis

The species energy hypothesis suggests the amount of available energy sets limits to the richness of the system. Thus, increased solar energy (with an abundance of water) at low latitudes causes increased net primary productivity (or photosynthesis). This hypothesis proposes the higher the net primary productivity the more individuals can be supported, and the more species there will be in an area. Put another way, this hypothesis suggests that extinction rates are reduced towards the equator as a result of the higher populations sustainable by the greater amount of available energy in the tropics. Lower extinction rates lead to more species in the tropics.

One critique of this hypothesis has been that increased species richness over broad spatial scales is not necessarily linked to increased number of individuals, which in turn is not necessarily related to increased productivity. Additionally, the observed changes in the number of individuals in an area with latitude or productivity are either too small (or in the wrong direction) to account for the observed changes in species richness. The potential mechanisms underlying the species-energy hypothesis, their unique predictions and empirical support have been assessed in a major review by Currie et al. (2004).

The effect of energy has been supported by several studies in terrestrial and marine taxa

Climate Harshness Hypothesis

Another climate-related hypothesis is the climate harshness hypothesis, which states the latitudinal diversity gradient may exist simply because fewer species can physiologically tolerate conditions at higher latitudes than at low latitudes because higher latitudes are often colder and drier than tropical latitudes. Currie et al. (2004) found fault with this hypothesis by stating that, although it is clear that climatic tolerance can limit species distributions, it appears that species are often absent from areas whose climate they can tolerate.

Climate Stability Hypothesis

Similarly to the climate harshness hypothesis, climate stability is suggested to be the reason for the latitudinal diversity gradient. The mechanism for this hypothesis is that while a fluctuating environment may increase the extinction rate or preclude specialization, a constant environment can allow species to specialize on predictable resources, allowing them to have narrower niches and facilitating speciation. The fact that temperate regions are more variable both seasonally and over geological timescales (discussed in more detail below) suggests that temperate regions are thus expected to have less species diversity than the tropics.

Critiques for this hypothesis include the fact that there are many exceptions to the assumption that climate stability means higher species diversity. For example, low species diversity is known to occur often in stable environments such as tropical mountaintops. Additionally, many habitats with high species diversity do experience seasonal climates, including many tropical regions that have highly seasonal rainfall (Brown and Lomolino 1998).

Historical/Evolutionary Hypotheses

There are three main hypotheses that are related to historical and evolutionary explanations for the increase of species diversity towards the equator.

The Historical Perturbation Hypothesis

The historical perturbation hypothesis proposes the low species richness of higher latitudes is a consequence of an insufficient time period available for species to colonize or recolonize areas because of historical perturbations such as glaciation (Brown and Lomolino 1998, Gaston and Blackburn 2000). This hypothesis suggests that diversity in the temperate regions have not yet reached equilibrium, and that the number of species in temperate areas will continue to increase until saturated (Clarke and Crame 2003).

The Evolutionary Rate Hypothesis

The evolutionary rate hypothesis argues higher evolutionary rates in the tropics have caused higher speciation rates and thus increased diversity at low latitudes (Cardillo et al. 2005, Weir & Schluter 2007, Rolland et al. 2014). Higher evolutionary rates in the tropics have been attributed to higher ambient temperatures, higher mutation rates, shorter generation time and/or faster physiological processes (Rohde 1992, Allen et al. 2006). Faster rates of microevolution in warm climates (i.e. low latitudes and altitudes) have been shown for plants (Wright et al. 2006), mammals (Gillman et al. 2009) and amphibians (Wright et al. 2010). Based on the expectation that faster rates of microevolution result in faster rates of speciation, these results suggest that faster evolutionary rates in warm climates almost certainly have a strong influence on the latitudinal diversity gradient. More research needs to be done to determine whether or not speciation rates actually are higher in the tropics. Understanding whether extinction rate varies with latitude will also be important to whether or not this hypothesis is supported (Rolland et al. 2014).

The Hypothesis of Effective Evolutionary Time

This hypothesis assumes that diversity is determined by the evolutionary time under which ecosystems have existed under relatively unchanged conditions, and by evolutionary speed directly determined by effects of environmental energy (temperature) on mutation rates, generation times, and speed of selection (Rohde 1992). It differs from most other hypotheses in not postulating an upper limit to species richness set by various abiotic and biotic factors, i.e., it is a nonequilibrium hypothesis assuming a largely non-saturated niche space. It does accept that many other factors may play a role in causing latitudinal gradients in species richness as well. The hypothesis is supported by much recent evidence, in particular the studies of Allen et al. (2006) and Wright et al. (2006).

Biotic Hypotheses

Biotic hypotheses claim ecological species interactions such as competition, predation, mutualism, and parasitism are stronger in the tropics and these interactions promote species coexistence and specialization of species, leading to greater speciation in the tropics. These hypotheses are problematic because they cannot be the proximate cause of the latitudinal diversity gradient as they fail to explain why species interactions might be stronger in the tropics. An example of one such hypothesis is the greater intensity of predation and more specialized predators in the tropics has contributed to the increase of diversity in the tropics (Pianka 1966). This intense predation could reduce the importance of competition (see competitive exclusion) and permit greater niche overlap and promote higher richness of prey. However, as discussed above, even if predation is more intense in the tropics (which is not certain), as it cannot be the ultimate cause of species diversity in the tropics because it fails to explain what gives rise to the richness of the predators in the tropics.

Several recent studies have failed to observe consistent changes in ecological interactions with latitude (Lambers et al. 2002, Hillebrand 2004). These studies suggest the intensity of species interactions are not correlated with the change in species richness with latitude.

Synthesis and Conclusions

There are many other hypotheses related to the latitudinal diversity gradient, but the above hypotheses are a good overview of the major ones still cited today. It is important to note that many of these hypotheses are similar to and dependent on one another. For example, the evolutionary hypotheses are closely dependent on the historical climate characteristics of the tropics.

The generality of the Latitudinal Diversity Gradient

Recently, Hillebrand (2004) performed an extensive meta-analysis of nearly 600 latitudinal gradients from published literature to determine the generality of the latitudinal diversity gradient across different organismal, habitat and regional characteristics. Hillebrand found the latitudinal gradient occurs in marine, terrestrial, and freshwater ecosystems, in both hemispheres. The gradient is steeper and more pronounced in richer taxa (i.e. taxa with more species), larger organisms, in marine and terrestrial versus freshwater ecosystems, and at regional versus local scales. The gradient steepness (the amount of change in species richness with latitude) is not influenced by dispersal, animal physiology (homeothermic or ectothermic) trophic level, hemisphere, or the latitudinal range of study. The study could not directly falsify or support any of the above hypotheses, however results do suggest a combination of energy/climate and area processes likely contribute to the latitudinal species gradient. Notable exceptions to the trend include the ichneumonidae, shorebirds, penguins, and freshwater zooplankton.

Data Robustness

One of the main assumptions about LDGs and patterns in species richness is that the underlying data (i.e. the lists of species at specific locations) are complete. However, this assumption is not met in most cases. For instance, diversity patterns for blood parasites of birds suggest higher diversity in tropical regions, however, the data may be skewed by undersampling in rich faunal areas such as Southeast Asia and South America. For marine fishes, which are among the most studied taxonomic groups, current lists of species are considerably incomplete for most of the world's oceans. At a 3° (approx. 350km2) spatial resolution, less than 1.8% of the world's oceans have above 80% of their fish fauna currently described.

CONCLUSION

The fundamental macroecological question that the latitudinal diversity gradient depends on is 'What causes patterns in species richness?'. Species richness ultimately depends on whatever proximate factors are found to affect processes of speciation, extinction, immigration, and emigration. While some ecologists continue to search for the ultimate primary mechanism that causes the latitudinal richness gradient, many ecologists suggest instead this ecological pattern is likely to be generated by several contributory mechanisms (Gaston and Blackburn 2000, Willig et al. 2003, Rahbek et al. 2007). For now the debate over the cause of the latitudinal diversity gradient will continue until a groundbreaking study provides conclusive evidence or there is general consensus that multiple factors contribute to the pattern.

PRIORITY EFFECT

In ecology, a **priority effect** is the impact that a particular species can have on community development due to prior arrival at a site.

There are two basic types: *inhibitory priority effects* occur when a species that arrives first at a site negatively impacts a species that arrives later by reducing the availability of space or resources. *Facilitative priority effects* occur when a species that arrives first at a site alters abiotic or biotic conditions in ways that positively impact a species arriving later. Priority effects are a central and pervasive element of ecological community development. These effects have important implications for natural systems as well as ecological restoration efforts.

Theoretical Foundation

Community Succession Theory

Early in the 20th century, Frederic Clements and other plant ecologists suggested that ecological communities develop in a linear, directional manner towards a final, stable end-point: the climax community. Clements indicated that a site›s climax community would reflect local climate. He conceptualized the climax community as a «superorganism» that followed a defined developmental sequence.

Early ecological succession theory maintained that the directional shifts from one stage of succession to the next were induced by the plants themselves. In this sense, succession theory implicitly recognized priority effects; the prior arrival of certain species had important impacts on future community composition. At the same time, the climax concept implied that species shifts were predetermined. A given species would always appear at the same point during the development of the climax community, and always have the same impact on community development.

This static view of priority effects remained essentially unchanged by the concept of patch dynamics, which was introduced by Alex Watt in 1947. Watt conceived of plant communities as dynamic «mechanisms» that followed predetermined succession cycles. Although Watt questioned the idea of a stable endpoint to community development, he seemed to agree with Clements that each particular species had a predetermined role to play in community development. They viewed succession as a process driven by facilitation, in which each species made local conditions more suitable for another species.

Individualistic Approach

In 1926, Henry Gleason presented an alternative hypothesis in which plants were conceptualized as individuals rather than components of a superorganism. Gleason suggested that the distribution of various species across the landscape reflected species-specific dispersal limitations and environmental requirements rather than predetermined associations among species. Gleason set the stage for future research on priority effects by explaining that initially identical ponds colonized by different species could develop through succession into very different communities. Thus, Gleason contested the idea of a predetermined climax community and recognized that different colonizing species could produce alternative trajectories of community development. Frank Egler (1954) built on Gleason›s hypothesis by developing the Initial Floristic Composition model to describe community development in abandoned agricultural fields. According to this model, the set of species present in a field immediately after abandonment had strong influences on community development and final community composition. Although rooted in succession theory, this approach foreshadowed the development of alternative stable state and community assembly theory.

Alternative Stable States

In the 1970s, it was suggested that natural communities could be characterized by multiple or alternative stable states. In accordance with the conclusions of Gleason and Egler, multiple stable state models indicated that the same environment could support several different combinations of species. Theorists argued that historical context could play a central role in determining which stable state would be present at any given time. Robert May explained, "If there is a unique stable state, historical accidents are unimportant; if there are many alternative locally stable states, historical accidents can be of overriding significance.»

Community Assembly Theory

The development of assembly theory followed from the emergence of alternative stable state theory. Assembly theory explains community development processes in the context of multiple stable states. It asks why a particular type of community developed when other stable community types were possible. In contrast to succession theory, assembly theory was developed largely by animal ecologists and explicitly incorporated historical context.

In one of the first models based on this theory, Jared Diamond (1975) developed quantitative "assembly rules" to predict avian community composition on an archipelago. Although the idea of deterministic community assembly quickly drew criticism, the assembly approach, which emphasized historical contingency and multiple stable states, continued to gain support. Drake (1991) used an assembly model to demonstrate that different community types would result from different sequences of species invasions. In Drake's model, early invaders had major impacts on the invasion success of species that arrived later. Other modeling studies suggested that priority effects may be especially important when invasion frequency is low enough to allow species to become established before replacement, or when other factors that could drive assembly (e.g., competition, abiotic stress) are relatively unimportant. In a 1999 review, Belyea and Lancaster described three basic determinants of community assembly: dispersal constraints, environmental constraints, and internal dynamics.They identified priority effects as a manifestation of the interaction between dispersal constraints and internal dynamics.

Empirical Evidence

On January 25, 2007, a search of the ISI Web of Science citation index using the key word "priority effect*" returned 65 ecology-related studies. The first study to explicitly mention priority effects was published in 1983. Although early research focused on animals and aquatic systems, more recent studies have begun to examine terrestrial and plant-based priority effects. Inhibitory priority effects have been documented more frequently than facilitative priority effects. Priority effects among species within the same trophic level and functional group, or guild, have been documented more frequently than effects across trophic levels or guilds. Studies indicate that both abiotic (e.g. resource availability) and biotic (e.g. predation) factors can affect the strength of priority effects.

Marine

Most of the earliest empirical evidence for priority effects came from studies on aquatic animals. Sutherland (1974) found that final community composition varied depending on the initial order of larval recruitment in a community of small marine organisms (sponges, tunicates, hydroids, and other species). Shulman (1983) and Almany (2003) found strong priority effects among coral reef fish. The former study found that prior establishment by a territorial damselfish reduced establishment rates of other fish. The authors also identified cross-trophic priority effects; prior establishment by a predator fish reduced establishment rates of prey fishes.

In the late 1980s, several studies examined priority effects in marine microcosms. Robinson and Dickerson (1987) found that priority effects were important in some cases, but suggested, «Being the first to invade a habitat does not guarantee success; there must be sufficient time for the early colonist to increase its population size for it to pre-empt further colonization.»Robinson and Edgemon (1988) later developed 54 communities of phytoplankton species by varying invasion order, rate, and timing. They found that although invasion order (priority effects) could explain a small fraction of the resulting variation in community composition, most of the variation was explained by changes in invasion rate and invasion timing. These studies indicate that priority effects may not be the only or the most important historical factor affecting the trajectory of community development.

In a striking example of cross-trophic priority effects, Hart (1992) found that priority effects explain the maintenance of two alternate stable states in stream ecosystems. While a macroalga is dominant in some patches, sessile grazers maintain a "lawn" of small microalgae in others. If the sessile grazers colonize a patch first, they exclude the macroalga, and vice versa.

Amphibian

In two of the most commonly cited empirical studies on priority effects, Alford and Wilbur documented inhibitory and facilitative priority effects among and toad larvae in experimental ponds. They found that hatchlings of a toad species (*Bufo americanus*) exhibited higher growth and survivorship when introduced to a pond before those of a frog species (*Rana sphenocephala*). The frog larvae, however, did best when introduced after the toad larvae. Thus, prior establishment by the toad species facilitated the frog species, while prior establishment by the frog species inhibited the toad species. Studies on tree frogs have also documented both types of priority effect. Morin (1987) also observed that priority effects became less important in the presence of a predatory salamander. He hypothesized that predation mediated priority effects by reducing competition between frog species. Studies on larval insects and frogs in water-filled tree holes and stumps found that abiotic factors such as space, resource availability, and toxin levels can also be important in mediating priority effects.

Terrestrial

Terrestrial studies on priority effects are rare. The aforementioned ISI Web of Science search retrieved only 19 studies on fully terrestrial organisms, and all of these studies were published during or after 1993. Most studies have focused on arthropods or grassland plant species. In a lab experiment, Shorrocks and Bingley (1994) showed that prior arrival increased survivorship for two species of fruit flies; each fly species had inhibitory impacts on the other. A 1996 field study on desert spiders by Ehmann and MacMahon showed that the presence of species from one spider guild reduced establishment of spiders from a different guild. More recently, Palmer (2003) demonstrated that priority effects allowed a competitively subordinate ant species to avoid exclusion by a competitively dominant species. If the competitively subordinate ants were able to colonize first, they altered their host tree's morphology in ways that made it less suitable for other ant species. This study was especially important because it was able to identify a mechanism driving observed priority effects.

With the exception of an early study exploring the facilitative effects of litter deposition, studies that explicitly addressed terrestrial plant priority effects began to appear in the literature around the year 2000. A study on two species of introduced grasses in Hawaiian woodlands found that the species with inferior competitive abilities may be able to persist through priority effects. At least three studies have come to similar conclusions about the coexistence of native and exotic grasses in California grassland ecosystems. If given time to establish, native species can successfully inhibit the establishment of exotics. Authors of the various studies attributed the prevalence of exotic grasses in California to the low seed production and relatively poor dispersal ability of native species.

Emerging Concepts

Long-term Implications: Convergence and Divergence

Although many studies have documented priority effects, the persistence of these effects over time often remains unclear. Young(2001) indicated that both convergence (in which "communities proceed towards a predisturbance state regardless of historical conditions") and divergence (in which historical factors continue to affect the long-term trajectory of community development) are present in nature. Among studies of priority effects, both trends seem to have been observed. Fukami (2005) argued that a community could be both convergent and divergent at different levels of community organization. The authors studied experimentally-assembled plant communities and found that while the identities of individual species remained unique across different community replicates, species traits generally became more similar.

Trophic Ecology

Some studies indicate that priority effects can occur across guilds or trophic levels. Such priority effects could have dramatic impacts on community composition and food web structure. Even intra-guild priority effects could have important consequences at multiple trophic levels if the affected species are associated with unique predator or prey species. Consider, for example, a plant species that is eaten by a host-specific herbivore. Priority effects that influence the ability of the plant species to establish would indirectly affect the establishment success of the associated herbivore. Theoretical models have described cyclical assembly dynamics in which species associated with different suites of predators are able to repeatedly replace one another.

Intraspecific Aggregation

In situations where two species are introduced at the same time, spatial aggregation of a species' propagules could cause priority effects by initially reducing interspecific competition. Aggregation during recruitment and establishment could allow inferior competitors to coexist with or even displace competitive dominants over the long-term. Several modeling efforts have begun to examine the implications of spatial priority effects for species coexistence. Rejmanek (2002) suggested that only 10 empirical studies examining intraspecific aggregation had been published by 2002.

Mechanisms and New Organisms

The literature on priority effects is currently growing in both depth and breadth. A few studies have begun to explore the mechanisms driving observed priority effects. Moreover, although past studies focused on a small subset of species, recent papers indicate that priority effects may be important for a wide range of organisms, including fungi, birds, lizards,and salamanders.

Ecological Restoration

Priority effects have important implications for ecological restoration. In many systems, information about priority effects can help practitioners identify cost-effective strategies for improving the survival and persistence of certain species, especially species of inferior competitive ability. For example, in a study on the restoration of native Californian grasses and forbs, Lulow (2004) found that forbs could not establish in plots where she had previously planted bunchgrasses. When bunchgrasses were added to plots where forbs had already been growing for a year, forbs were able to coexist with grasses for at least 3–4 years. Lulow's results suggested that planting forbs before grasses might improve forb persistence in this system.

REFERENCES

1. Young, Truman P.; Chase, Jonathan M.; Huddleston, Russell T. (2001-03-20). "Community Succession and Assembly Comparing, Contrasting and Combining

Paradigms in the Context of" (PDF). Ecological Restoration. 19 (1): 5–18. doi:10.3368/er.19.1.5. ISSN 1543-4079.

2. Lulow, Megan Elizabeth (2004). Restoration of California's Inland Grasslands: The Role of Priority Effects and Management Strategies in Establishing Native Communities and the Ability of Native Grasses to Resist Invasion by Non-native Grasses (Ph.D.). University of California, Davis.
3. ^ Jump up to:a b c d Fukami, Tadashi; Martijn Bezemer, T.; Mortimer, Simon R.; van der Putten, Wim H. (2005-12-01). "Species divergence and trait convergence in experimental plant community assembly" (PDF). Ecology Letters. 8 (12): 1283–1290. doi:10.1111/j.1461-0248.2005.00829.x. ISSN 1461-0248.
4. Connell, Joseph H; Slatyer, Ralph O (1977). "Mechanisms of succession in natural communities and their role in community stability and organization". American naturalist. 111: 1119–1144. doi:10.1086/283241. ISSN 0003-0147. JSTOR 2460259.
5. b Young, T. P.; Petersen, D. A.; Clary, J. J. (2005-06-01). "The ecology of restoration: historical links, emerging issues and unexplored realms" (PDF). Ecology Letters. 8 (6): 662–673. doi:10.1111/j.1461-0248.2005.00764.x. ISSN 1461-0248.
6. Clements, Frederic E. (1936-02). "Nature and Structure of the Climax" (PDF). The Journal of Ecology. 24 (1): 252. doi:10.2307/2256278. ISSN 0022-0477. JSTOR 2256278. Check date values in: |date= (help)
7. Clements, Frederic Edward (1916). Plant Succession: An Analysis of the Development of Vegetation. Carnegie Institution of Washington.
8. Tansley, A. G. (1935-07). "The Use and Abuse of Vegetational Concepts and Terms" (PDF). Ecology. 16 (3): 284–307. doi:10.2307/1930070. ISSN 0012-9658. JSTOR 1930070. Check date values in: |date= (help)
9. Watt, Alex S. (1947-12). "Pattern and Process in the Plant Community" (PDF). The Journal of Ecology. 35 (1/2): 1–22. doi:10.2307/2256497. ISSN 0022-0477. JSTOR 2256497. Check date values in: |date= (help)
10. Gleason, H. A. (1926-01). "The Individualistic Concept of the Plant Association" (PDF). Bulletin of the Torrey Botanical Club. 53 (1): 7. doi:10.2307/2479933. ISSN 0040-9618. Check date values in: |date= (help)
11. Egler, Frank E. (1954-11-01). "Vegetation science concepts I. Initial floristic composition, a factor in old-field vegetation development with 2 figs.". Vegetatio. 4 (6): 412–417. doi:10.1007/BF00275587. ISSN 1573-5052.
12. Lewontin, R C (1969). "The meaning of stability". Brookhaven symposia in biology. 22: 13–24. ISSN 0068-2799. PMID 5372787.
13. Holling, Crawford S (1973). "Resilience and stability of ecological systems". Annual Review of Ecology and Systematics: 1–23. ISSN 0066-4162. JSTOR 2096802.
14. May, Robert M (1977). "Thresholds and breakpoints in ecosystems with a multiplicity of stable states". Nature. 269 (5628): 471–477. doi:10.1038/269471a0.

15. Diamond, Jared M. (1975-01-01). "Assembly of species communities". In Cody, Martin L.; Diamond, Jared M. Ecology and Evolution of Communities. Harvard University Press. pp. 342–444. ISBN 9780674224445.

16. Connor, Edward F.; Simberloff, Daniel (1979-12). "The Assembly of Species Communities: Chance or Competition?". Ecology. 60 (6): 1132. doi:10.2307/1936961. ISSN 0012-9658. JSTOR 1936961. Check date values in: |date=(help)

17. Hughes, Terence P. (1989-02). "Community Structure and Diversity of Coral Reefs: The Role of History" (PDF). Ecology. 70 (1): 275. doi:10.2307/1938434. ISSN 0012-9658. Check date values in: |date= (help)

18. Drake, James A (1991). "Community-assembly mechanics and the structure of an experimental species ensemble". American Naturalist. 137: 1–26. doi:10.1086/285143. ISSN 0003-0147. JSTOR 2462154.

19. Lockwood, Julie L.; Powell, Robert D.; Nott, M. Philip; Pimm, Stuart L. (1997-12). "Assembling Ecological Communities in Time and Space" (PDF). Oikos. 80 (3): 549. doi:10.2307/3546628. ISSN 0030-1299. JSTOR 3546628. Check date values in: |date= (help)

20. Weiher, E.; Keddy, P. A. (1995). "Assembly rules, null models, and trait dispersion: new questions front old patterns". Oikos. 74: 159–164. doi:10.2307/3545686.

21. Belyea, Lisa R.; Lancaster, Jill (1999-09). "Assembly Rules within a Contingent Ecology". Oikos. 86 (3): 402. doi:10.2307/3546646. ISSN 0030-1299. JSTOR 3546646. Check date values in: |date= (help)

22. Sutherland, John P (1974). "Multiple stable points in natural communities". American Naturalist. 108 (964): 859–873. doi:10.1086/282961. ISSN 0003-0147. JSTOR 2459615.

23. Shulman, Myra J.; Ogden, John C.; Ebersole, John P.; McFarland, William N.; Miller, Steven L.; Wolf, Nancy G. (1983-12-01). "Priority Effects in the Recruitment of Juvenile Coral Reef Fishes". Ecology. 64 (6): 1508–1513. doi:10.2307/1937505. ISSN 0012-9658.

24. Almany, Glenn R. (2003-07-01). "Priority Effects in Coral Reef Fish Communities" (PDF). Ecology. 84 (7): 1920–1935. doi:10.1890/0012-9658(2003)084[1920:PEICRF]2.0.CO;2. ISSN 0012-9658.

25. Robinson, James F.; Dickerson, Jaime E. (1987-06-01). "Does Invasion Sequence Affect Community Structure?". Ecology. 68 (3): 587–595. doi:10.2307/1938464. ISSN 0012-9658.

26. Robinson, James V.; Edgemon, Michael A. (1988-10). "An Experimental Evaluation of the Effect of Invasion History on Community Structure". Ecology. 69 (5): 1410–1417. doi:10.2307/1941638. ISSN 0012-9658. Check date values in: |date= (help)

27. c Hart, David D. (1992-08-01). "Community organization in streams: the importance of species interactions, physical factors, and chance" (PDF). Oecologia. 91 (2): 220–228. doi:10.1007/BF00317787. ISSN 1432-1939.

28. Alford, Ross A.; Wilbur, Henry M. (1985-08-01). "Priority Effects in Experimental Pond Communities: Competition between Bufo and Rana". Ecology. 66 (4): 1097–1105. doi:10.2307/1939161. ISSN 0012-9658. JSTOR 1939161.
29. Wilbur, Henry M.; Alford, Ross A. (1985-08). "Priority Effects in Experimental Pond Communities: Responses of Hyla to Bufo and Rana". Ecology. 66 (4): 1106–1114. doi:10.2307/1939162. ISSN 0012-9658. Check date values in: |date= (help)
30. Morin, Peter Jay (1987-06-01). "Predation, Breeding Asynchrony, and the Outcome of Competition Among Treefrog Tadpoles". Ecology. 68 (3): 675–683. doi:10.2307/1938473. ISSN 0012-9658.
31. Warner, Susan C.; Dunson, William A.; Travis, Joseph (1991-11-01). "Interaction of pH, density, and priority effects on the survivorship and growth of two species of hylid tadpoles". Oecologia. 88 (3): 331–339. doi:10.1007/BF00317575. ISSN 1432-1939.
32. Fincke, OlA. M. (1999-02-01). "Organization of predator assemblages in Neotropical tree holes: effects of abiotic factors and priority" (PDF). Ecological Entomology. 24 (1): 13–23. doi:10.1046/j.1365-2311.1999.00166.x. ISSN 1365-2311.
33. Sunahara, Toshihiko; Mogi, Motoyoshi (2002-06-01). "Priority effects of bamboo-stump mosquito larvae: influences of water exchange and leaf litter input". Ecological Entomology. 27 (3): 346–354. doi:10.1046/j.1365-2311.2002.00417.x. ISSN 1365-2311.
34. Shorrocks, B.; Bingley, M. (1994). "Priority effects and species coexistence: experiments with fungal-breeding Drosophila". Journal of animal ecology. 63 (4): 799–806. doi:10.2307/5257. ISSN 0021-8790.
35. Ehmann, William J; MacMahon, James A (1996). "Initial tests for priority effects among spiders that co-occur on sagebrush shrubs". Journal of Arachnology. 24 (3): 173–185. ISSN 0161-8202.
36. b Palmer, Todd M.; Stanton, Maureen L.; Young, Truman P. (2003-10). "Competition and Coexistence: Exploring Mechanisms That Restrict and Maintain Diversity within Mutualist Guilds". The American Naturalist. 162 (s4): S63–S79. doi:10.1086/378682. ISSN 1537-5323. JSTOR 10.1086/378682. Check date values in: |date=(help)
37. Facelli, J. M.; Facelli, E. (1993-08-01). "Interactions after death: plant litter controls priority effects in a successional plant community" (PDF). Oecologia. 95 (2): 277–282. doi:10.1007/BF00323500. ISSN 1432-1939.
38. D'Antonio, Carla M.; Hughes, R. Flint; Vitousek, Peter M. (2001-01-01). "Factors Influencing Dynamics of two Invasive C4 Grasses in Seasonally Dry Hawaiian Woodlands". Ecology. 82 (1): 89–104. doi:10.1890/0012-9658(2001)082[0089:FIDOTI]2.0.CO;2. ISSN 0012-9658.
39. Seabloom, Eric W.; Harpole, W. Stanley; Reichman, O. J.; Tilman, David (2003-11-11). "Invasion, competitive dominance, and resource use by exotic and native California grassland species". Proceedings of the National Academy of Sciences.

100 (23): 13384–13389. doi:10.1073/pnas.1835728100. ISSN 1091-6490. PMC 263823 . PMID 14595028.

40. Corbin, Jeffrey D.; D'Antonio, Carla M. (2004-05-01). "Competition Between Native Perennial and Exotic Annual Grasses: Implications for a Historical Invasion" (PDF). Ecology. 85 (5): 1273–1283. doi:10.1890/02-0744. ISSN 0012-9658.
41. Lulow, Megan E. (2006-12-01). "Invasion by Non-Native Annual Grasses: The Importance of Species Biomass, Composition, and Time Among California Native Grasses of the Central Valley" (PDF). Restoration Ecology. 14 (4): 616–626. doi:10.1111/j.1526-100X.2006.00173.x. ISSN 1526-100X.
42. Leibold, M. A.; Holyoak, M.; Mouquet, N.; Amarasekare, P.; Chase, J. M.; Hoopes, M. F.; Holt, R. D.; Shurin, J. B.; Law, R.; Tilman, D.; Loreau, M.; Gonzalez, A. (2004-07-01). "The metacommunity concept: a framework for multi-scale community ecology". Ecology Letters. 7 (7): 601–613. doi:10.1111/j.1461-0248.2004.00608.x. ISSN 1461-0248.
43. Steiner, Christopher F.; Leibold, Mathew A. (2004-01-01). "Cyclic assembly trajectories and scale-dependent productivity–diversity relationships" (PDF). Ecology. 85 (1): 107–113. doi:10.1890/03-3010. ISSN 0012-9658.
44. Inouye, Brian D. (1999-01-01). "Integrating nested spatial scales: implications for the coexistence of competitors on a patchy resource". Journal of Animal Ecology. 68 (1): 150–162. doi:10.1046/j.1365-2656.1999.00272.x. ISSN 1365-2656.
45. Chesson, Peter (2000). "Mechanisms of Maintenance of Species Diversity". Annual Review of Ecology and Systematics. 31 (1): 343–366. doi:10.1146/annurev.ecolsys.31.1.343.
46. Hartley, Stephen; Shorrocks, Bryan (2002-07-01). "A general framework for the aggregation model of coexistence". Journal of Animal Ecology. 71 (4): 651–662. doi:10.1046/j.1365-2656.2002.00628.x. ISSN 1365-2656.
47. Molofsky, Jane; Bever, James D. (2002-12-07). "A novel theory to explain species diversity in landscapes: positive frequency dependence and habitat suitability". Proceedings of the Royal Society B: Biological Sciences. 269 (1508): 2389–2393. doi:10.1098/rspb.2002.2164. ISSN 1471-2954. PMC 1691177 . PMID 12495479.
48. Rejmanek, M. (2002). "Intraspecific aggregation and species coexistence". Trends in Ecology & Evolution. 17: 209–210. doi:10.1016/s0169-5347(02)02494-1.
49. Kennedy, Peter G.; Bruns, Thomas D. (2005-05-01). "Priority effects determine the outcome of ectomycorrhizal competition between two Rhizopogon species colonizing Pinus muricata seedlings". New Phytologist. 166 (2): 631–638. doi:10.1111/j.1469-8137.2005.01355.x. ISSN 1469-8137.
50. Rohlfs, Marko (2005-09-01). "Density-dependent insect-mold interactions: effects on fungal growth and spore production". Mycologia. 97 (5): 996–1001. doi:10.3852/mycologia.97.5.996. ISSN 1557-2536. PMID 16596951.
51. Gamarra, Javier G. P.; Montoya, José M.; Alonso, David; Solé, Ricard V. (2005-03-01). "Competition and introduction regime shape exotic bird communities

in Hawaii" (PDF). Biological Invasions. 7 (2): 297–307. doi:10.1007/s10530-004-0876-3. ISSN 1573-1464.

52. M'Closkey, Robert T.; Hecnar, Stephen J.; Chalcraft, David R.; Cotter, Jill E. (1998-10-01). "Size distributions and sex ratios of colonizing lizards" (PDF). Oecologia. 116 (4): 501–509. doi:10.1007/s004420050615. ISSN 1432-1939.
53. Eitam, Avi; Blaustein, Leon; Mangel, Marc (2005-11-01). "Density and intercohort priority effects on larval Salamandra salamandra in temporary pools". Oecologia. 146 (1): 36–42. CiteSeerX 10.1.1.118.8240 . doi:10.1007/s00442-005-0185-2. ISSN 0029-8549.
54. Suding, Katharine N.; Gross, Katherine L.; Houseman, Gregory R. (2004-01-01). "Alternative states and positive feedbacks in restoration ecology" (PDF). Trends in Ecology & Evolution. 19 (1): 46–53. doi:10.1016/j.tree.2003.10.005. ISSN 0169-5347. PMID 16701225.
55. Dov Sax, Aug 2008. The Proceedings of the National Academy of Sciences Archived July 6, 2012, at the Wayback Machine.
56. Carlton, James T. 2002. Introduced Species in U.S. Coastal Waters. Pew Oceans Commission.
57. U.S. Environmental Protection Agency. Mid-Atlantic Integrated Assessment. September 16, 2003. Introduced species. Web site at US EPA
58. Richardson et al. 2000[dead link]
59. biodiversity[dead link]
60. Colautti and MacIsaac 2004[dead link]
61. CEQ (1999). Web site page with Executive Order 13112text.
62. 2013. Invasive Species Program. Clemson University - DPI. http://www.clemson.edu/public/regulatory/plant_industry/invasive_exotic_programs/. Accessed 24 May 2013.
63. Naylor, R.L., S.L. Williams, and D.R. Strong. 2001. Aquaculture—a gateway for exotic species. Science 294: 1655-56.
64. Riley, SPD, H.B. Shaffer, S.R. Voss, B.M. Fitzpatrick. Hybridization between a rare, native tiger salamander and its introduced congener. 2003. Ecological Applications 13: 1263–1275.
65. Foster, J. and A. Sandberg. Friends or foe? Invasive species and public green space in Toronto. 2004. The Geographical Review 94: 178-198.
66. Deichsel, G. & Gist, D. H. On the Origin of the Common Wall Lizards Podarcis muralis (Reptilia: Lacertidae) in Cincinnati, Ohio
67. Shirey, P.D. and G.A. Lamberti. 2010. Assisted colonization under the U.S. Endangered Species Act. Conservation Letters 3(1): 45-52.
68. Moritz, Craig. 1999. Conservation units and translocations: Strategies for conserving evolutionary processes. Hereditas 130: 217-228.
69. David Pimentel, Lori Lach, Rodolfo Zuniga, and Doug Morrison, Environmental and Economic Costs Associated with Non-Indigenous Species in the United

States, College of Agriculture and Life Sciences, Cornell University (Ithaca, New York), June 12, 1999.

70. Corn; Tim Johnson, “Invasive Species,” The Burlington Free Press, November 9, 2003

71. Cohen, A.N. And J.T. Carlton. 1998. Accelerating invasion rate in a highly invaded estuary. Science 279: 555-558.

72. Kennedy, K. 1992. A systematic study of the genus Phyla Lour (Verbenaceae: Verbenoideae, Lantanae). University of Texas at Austin, Austin, Texas, USA.

73. Xu, Cheng-Yuan; Tang, Shaoqing; Fatemi, Mohammad; Gross, Caroline L.; Julien, Mic H.; Curtis, Caitlin; van Klinken, Rieks D. (2015-09-01). “Population structure and genetic diversity of invasive Phyla canescens: implications for the evolutionary potential”. Ecosphere. 6 (9): 1–21. doi:10.1890/ES14-00374.1. ISSN 2150-8925.

74. Whalley, R. D. B.; Price, J. N.; Macdonald, M. J.; Berney, P. J. “Drivers of change in the Social-Ecological Systems of the Gwydir Wetlands and Macquarie Marshes in northern New South Wales, Australia”. The Rangeland Journal. 33 (2). doi:10.1071/rj11002.

75. Olivier, L., J. P. Galland,and H. Maurin, editors.1995. Livre Rouge de la flore menacée de France.Tome I. Espèces prioritaires. SPN-IEGB /MNHN, DNP/ Ministère Environnement, CBN Porquerolles, Paris, France

76. 2013. “Pest Alert: Citrus Greening.” Clemson University Invasive Species Program. http://www.clemson.edu/public/regulatory/plant_industry/invasive_exotic_programs/Pest%20Alerts/citrus_greening.html. Accessed 24 May 2013.

77. Assessment of Planetary Protection and Contamination Control Technologies for Future Planetary Science Missions, Jet Propulsion Laboratory, January 24, 2011

78. Sahney, S.; Benton, M.J. (2008). “Recovery from the most profound mass extinction of all time” (PDF). Proceedings of the Royal Society B. 275 (1636): 759–65. doi:10.1098/rspb.2007.1370. PMC 2596898 . PMID 18198148.

79. Mora C & Robertson DR; Robertson (2005). “Causes of latitudinal gradients in species richness: a test with fishes of the Tropical Eastern Pacific” (PDF). Ecology. 86 (7): 1771–1792. doi:10.1890/04-0883.

80. Dowle, E. J., M. Morgan-Richards, S. A. Trewick. 2013 Molecular evolution and the latitudinal biodiversity gradient. Heredity 110:501-510 doi:10.1038/hdy.2013.4.

81. Tittensor D.; et al. (2011). “Global patterns and predictors of marine biodiversity across taxa” (PDF). Nature. 466 (7310): 1098–1101. doi:10.1038/nature09329. PMID 20668450.

82. Mora C; et al. (2003). “Patterns and processes in reef fish diversity” (PDF). Nature. 421 (6926): 933–936. doi:10.1038/nature01393. PMID 12606998.

83. Cardillo, M.; Orme, C. D. L.; Owens, I. P. F. (2005). “Testing for latitudinal bias in diversification rates: An example using New World birds”. Ecology. 86 (9): 2278–2287. doi:10.1890/05-0112.

84. Currie, D. J.; Mittelbach, G. G.; Cornell, H. V.; Kaufman, D. M.; Kerr, J. T.; Oberdorff, T. (2004). "Predictions and tests of climate-based hypotheses of broad-scale variation in taxonomic richness". Ecology Letters. 7 (11): 1121–1134. doi:10.1111/j.1461-0248.2004.00671.x.

85. Clark, Nicholas; Clegg, S.; Lima, M. (2014). "A review of global diversity in avian haemosporidians (Plasmodium and Haemoproteus: Haemosporida): new insights from molecular data". International Journal for Parasitology. 44 (5): 329–338. doi:10.1016/j.ijpara.2014.01.004. PMID 24556563.

86. Mora C; et al. (2007). "The completeness of taxonomic inventories for describing the global diversity and distribution of marine fishes" (PDF). Proceedings of the Royal Society B. 275 (1631): 149–155. doi:10.1098/rspb.2007.1315.

87. Allen, A. P.; Gillooly, J. F.; Savage, V. M.; Brown, J. H. (2006). "Kinetic effects of temperature on rates of genetic divergence and speciation". PNAS. 103 (24): 9130–9135. doi:10.1073/pnas.0603587103. PMC 1474011 . PMID 16754845.

88. Blackburn, T. M.; Gaston, K. J. (1997). "The relationship between geographic area and the latitudinal gradient in species richness in New World birds". Evolutionary Ecology. 11: 195–204. doi:10.1023/A:1018451916734.

89. Bokma, F. J.; Monkkonen, M.; Mönkkönen, Mikko (2001). "Random processes and geographic species richness patterns: why so few species in the north?". Ecography. 24: 43–49. doi:10.1034/j.1600-0587.2001.240106.x.

90. Brown, J. H., and M. V. Lomolino. 1998. Biogeography. Sinauer Associates, Sunderland.

91. Cardillo, M.; Orme, C. D. L.; Owens, I. P. F. (2005). "Testing for latitudinal bias in diversification rates: An example using New World birds". Ecology. 86 (9): 2278–2287. doi:10.1890/05-0112.

92. Clarke, A., and J. A. Crame. 2003. The importance of historical processes in global patterns of diversity. Pages 130-151 in T. M. Blackburn and K. J. Gaston, editors. Macroecology Concepts and Consequences. Blackwell Scientific, Oxford.

93. Colwell, R. K.; Hurtt, G. C. (1994). "Nonbiological gradients in species richness and a spurious Rapoport effect". American Naturalist. 144 (4): 570–595. doi:10.1086/285695.

94. Colwell, R. K.; Lees, D. C. (2000). "The mid-domain effect: geometric constraints on the geography of species richness". Trends in Ecology & Evolution. 15 (2): 70–76. doi:10.1016/s0169-5347(99)01767-x.

95. Colwell, R. K.; Rahbek, C.; Gotelli, N. (2004). "The mid-domain effect and species richness patterns: what have we learned so far?". American Naturalist. 163: E1–E23. doi:10.1086/382056.

96. Colwell, R. K.; Rahbek, C.; Gotelli, N. (2005). "The mid-domain effect: there's a baby in the bathwater". American Naturalist. 166: E149–E154. doi:10.1086/491689.

97. Connolly, S. R. (2005). "Process-based models of species distributions and the mid-domain effect". American Naturalist. 166 (1): 1–11. doi:10.1086/430638.

PMID 15937785.

98. Crame, J. A. (2001). "Taxonomic diversity gradients through geological time". Diversity and Distributions. 7 (4): 175–189. doi:10.1046/j.1472-4642.2001.00106.x.

99. Currie, D. J.; Kerr, J. T. (2007). "Testing, as opposed to supporting, the Mid-domain Hypothesis: a response to Lees and Colwell (2007)". Ecology Letters. 10: E9–E10. doi:10.1111/j.1461-0248.2007.01074.x.

100. Currie, D. J.; Mittelbach, G. G.; Cornell, H. V.; Kaufman, D. M.; Kerr, J. T.; Oberdorff, T. (2004). "Predictions and tests of climate-based hypotheses of broad-scale variation in taxonomic richness". Ecology Letters. 7: 1121–1134. doi:10.1111/j.1461-0248.2004.00671.x.

101. Diniz-Filho, J. A.; Rangel, T. F. L. V. B.; De Souza, Marcia Christianne; Rangel, Thiago F. L. V. B. (2002). "Null models and spatial patterns of species richness in South American birds of prey". Ecology Letters. 5: 47–55. doi:10.1046/j.1461-0248.2002.00289.x.

102. Dunn, R. R.; McCain, C. M.; Sanders, N. (2007). "When does diversity fit null model predictions? Scale and range size mediate the mid-domain effect.". Global Ecology & Biogeography. 3: 305–312.

103. Gaston, K. J. (2000). "Global patterns in biodiversity". Nature. 405 (6783): 220–227. doi:10.1038/35012228. PMID 10821282.

104. Gaston, K. J., and T. M. Blackburn. 2000. Pattern and processes in macroecology. Blackwell Scientific, Oxford.

105. Hawkins, B. A.; Diniz-Filho, J. A. F. (2002). "The mid-domain effect cannot explain the diversity gradient of Nearctic birds". Global Ecology and Biogeography. 11 (5): 419–426. doi:10.1046/j.1466-822x.2002.00299.x.

106. Gillman, L. N.; Keeling, D. J.; Ross, H. A.; Wright, S. D. (2009). "Latitude, elevation and the tempo of molecular evolution in mammals". Proceedings of the Royal Society B: Biological Sciences. 276 (1671): 3353–3359. doi:10.1098/rspb.2009.0674.

107. Hawkins, B. A.; Diniz-Filho, J. A. F.; Weis, A. E. (2005). "The mid-domain effect and diversity gradients: is there anything to learn". American Naturalist. 166: E140–E143. doi:10.1086/491686.

108. Hawkins, B. A.; Porter, E. E. (2001). "Area and the latitudinal diversity gradient for terrestrial birds". Ecology Letters. 4 (6): 595–601. doi:10.1046/j.1461-0248.2001.00271.x.

109. Hillebrand, H. (2004). "On the generality of the latitudinal diversity gradient". The American Naturalist. 163 (2): 192–211. doi:10.1086/381004. PMID 14970922.

110. Jetz, W.; Rahbek, C. (2001). "Geometric constraints explain much of the species richness pattern in African birds". Proceedings of the National Academy of Sciences of the USA. 98 (10): 5661–5666. doi:10.1073/pnas.091100998. PMID 11344307.

111. Kerr, J. T.; Perring, M.; Currie, D. J. (2006). "The missing Madagascan mid-domain effect". Ecology Letters. 9 (2): 149–159. doi:10.1111/j.1461-0248.2005.00860.x.

PMID 16958880.

112. Koleff, P.; Gaston, K. J. (2001). "Latitudinal gradients in diversity: real patterns and random models". Ecography. 24 (3): 341–351. doi:10.1111/j.1600-0587.2001.tb00207.x.

113. Lambers, J. H. R.; Clark, J. S.; Beckage, B. (2002). "Density-dependent mortality and the latitudinal gradient in species diversity". Nature. 417 (6890): 732–735. doi:10.1038/nature00809. PMID 12066182.

114. Lees, D. C.; Colwell, R. K. (2007). "A strong Madagascan rainforest MDE and no equatorward increase in species richness: Re-analysis of 'The missing Madagascan mid-domain effect', by Kerr J.T., Perring M. & Currie D.J (Ecology Letters 9:149-159, 2006)". Ecology Letters. 10: E4–E8. doi:10.1111/j.1461-0248.2007.01040.x.

115. Pianka, E. R. (1966). "Latitudinal gradients in species diversity: a review of concepts". American Naturalist. 100 (910): 33–46. doi:10.1086/282398.

116. Pimm, S.L.; Brown, J.H. (2004). "Domains of diversity". Science. 304 (5672): 831–833. doi:10.1126/science.1095332. PMID 15131295.

117. Rahbek, C.; Gotelli, N.; Colwell, R.K.; Entsminger, G.L.; Rangel, T.F.L.V.B.; Graves, G.R. (2007). "Predicting continental-scale patterns of bird species richness with spatially explicit models". Proceedings of the Royal Society of London Series B. 274 (1607): 165–174. doi:10.1098/rspb.2006.3700. PMID 17148246.

118. Rohde, K. (1992). "Latitudinal gradients in species diversity: the search for the primary cause". Oikos. 65 (3): 514–527. doi:10.2307/3545569. JSTOR 3545569.

119. Rohde, K. (1997). "The larger area of the tropics does not explain latitudinal gradients in species diversity". Oikos. 79: 169–172. doi:10.2307/3546102. JSTOR 3546102.

120. Rolland, J.; Condamine, F.L.; Jiguet, F.; Morlon, H. (2014). "Faster speciation and reduced extinction in the tropics contribute to the mammalian latitudinal diversity gradient". PLoS Biology. 12: e1001775. doi:10.1371/journal.pbio.1001775.

121. Romdal, T. S.; Colwell, R. K.; Rahbek, C. (2005). "The influence of band sum area, domain extent and range sizes on the latitudinal mid-domain effect". Ecology. 86: 235–244. doi:10.1890/04-0096.

122. Rosenzweig, M. L. (1992). "Species diversity gradients: we know more and less than we thought". Journal of Mammalogy. 73 (4): 715–730. doi:10.2307/1382191. JSTOR 1382191.

123. Rosenzweig, M. L. 2003. How to reject the area hypothesis of latitudinal gradients. Pages 87–106 in T. M. Blackburn and K. J. Gaston, editors. Macroecology: Concepts and Consequences. Blackwell Publishing, Oxford.

124. Storch, D.; Davies, R.G.; Zajicek, S.; Orme, C.D.L.; Olson, V.; Thomas, G.H.; Ding, T.S.; Rasmussen, P.C.; Ridgely, R.S.; et al. (2006). "Energy, range dynamics and global species richness patterns: reconciling mid-domain effects and environmental determinants of avian diversity". Ecology Letters. 9 (12): 1308–1320. doi:10.1111/j.1461-0248.2006.00984.x. PMID 17118005.

125. Terborgh, J. (1973). "On the notion of favorableness in plant ecology". American Naturalist. 107 (956): 481–501. doi:10.1086/282852.

126. Weir, J.T.; Schluter, D. (2007). "The latitudinal gradient in recent speciation and extinction rates of birds and mammals". Science. 315 (5818): 1574–1576. doi:10.1126/science.1135590. PMID 17363673.

127. Willig, M. R.; Lyons, S. K. (1998). "An analytical model of latitudinal gradients of species richness with an empirical test for marsupials and bats in the New World". Oikos. 81: 93–98. doi:10.2307/3546471. JSTOR 3546471.

128. Willig, M. R.; Kaufmann, D. M.; Stevens, R. D. (2003). "Latitudinal gradients of biodiversity: pattern, process, scale and synthesis". Annu. Rev. Ecol. Syst. 34: 273–309. doi:10.1146/annurev.ecolsys.34.012103.144032.

129. Wright, S.; Keeling, J.; Gillman, L. (2006). "The road from Santa Rosalia: a faster tempo of evolution in tropical climates". Proceedings of the National Academy of Sciences. 103 (20): 7718–7722. doi:10.1073/pnas.0510383103.

130. Zapata, F. A.; Gaston, K. J.; Chown, S. L. (2003). "Mid-domain models of species richness gradients: assumptions, methods and evidence". J. Animal Ecology. 72 (4): 677–690. doi:10.1046/j.1365-2656.2003.00741.x.

131. Zapata, F. A.; Gaston, K. J.; Chown, S. L. (2005). "The mid-domain effect revisited". American Naturalist. 166 (5): E144–E148. doi:10.1086/491685. PMID 16224717.

Index